The Wonderful World of Butterflies and Moths

The Wonderful World of Butterflies and Moths

Robert Goodden

Hamlyn
London · New York · Sydney · Toronto

Acknowledgements

Ardea Photographics: I. R. Beames, 28, 29T, 48L. J. L. Mason, 65. **Biofotos**: Heather Angel, 18T, 31TL. **Bruce Coleman Limited**: Jane Burton, 14, 22, 25TL, 67, 76B. A. J. Deane, 79. Inigo Everson, 38B. C. B. Frith, 72. John Markham, 34C. H. Rivarola, 23. **Worldwide Butterflies Limited**: Robert Goodden, 7T&B, 8, 9T&BL, 11, 12, 13T&B, 15, 16, 17T&B, 18B, 19T&B, 20–21, 24T&B, 25B, 26TL&R, 26B, 27, 29B, 30–31, 31B, 32, 33T&B, 34T, 37T, 38T, 39, 40T, 41, 42–43, 43T, 44, 45, 47, 48R, 49TL, 49TR, 49B, 50, 52, 53, 54B, 55, 57, 58T&B, 59, 60, 61, 62B, 63T&B, 64, 68, 69, 70T&B, 71, 73, 74, 75L, 76T, 80, 81, 83, 84, 85, 87, 88T, 89, 90–91, 91T&B, 92, 93T, 94. **Natural History Photographic Agency**: Anthony Bannister, title-page, 10, 31TR, 54T. F. Baillie, 35, 36. Stephen Dalton, *front jacket, front jacket flap, back jacket, back jacket flap,* endpapers, 6, 20, 25TR, 37B, 62T 66, 75R, 77, 82T, 88B, 90, 93B. Michael Davies 86B. G. E. Hyde, 9BR, 40B, 46, 82B. W. Zepf, 51, 56. **Wildlife Studies Limited**: 86T.

Published by
The Hamlyn Publishing Group Limited
London · New York · Sydney · Toronto
Astronaut House, Feltham, Middlesex,
England

ISBN 0 600 31392 1

Filmset by Filmtype Services Limited,
Scarborough, England
Printed in Spain by Mateu Cromo, Madrid

Contents

From egg to pupa

Above
The bright yellow eggs of the Black-veined White (*Aporia crataegi*) are laid in a closely arranged batch. The larvae live gregariously, forming a tight web in which they hibernate. Even in spring, after hibernation, they live together and will sometimes pupate in groups.

There are four stages in the life cycle of butterflies and moths, the Lepidoptera, and the stages form what is known as a complete metamorphosis. (Certain other insects hatch from the egg as miniatures of the adult and therefore have fewer stages and their changes are known as incomplete metamorphosis.) The caterpillar hatches from the egg, grows and forms a chrysalis, from which the adult emerges to breed and produce further eggs. Large numbers are usually laid and there is a tremendous loss of individuals at all stages. Predators take quite a share of a butterfly's brood and all along the line there are losses through disease or unfavourable conditions. This is all a part of the natural food chain and balance of nature and if all the brood were to survive the world would become a seething mass of insects! The duration of each stage is dependent on climate and the species. A tropical species might be only three days in the egg stage, a caterpillar for eight days, a pupa for seven days – just eighteen days before it emerges as a butterfly or moth. In a temperate climate a fast-growing species would take about eight weeks. A number of species complete the life cycle only once a year, spending much of their time in hibernation.

Eggs of butterflies and moths are quite distinctive, enabling us to recognize the species almost as easily as by seeing the adult insect. They are worth examining under a microscope and under a binocular instrument they are particularly striking. Eggs of Pieridae are similar in shape to a milk bottle, beautifully ribbed and engraved. *Papilio* species lay eggs that are almost perfectly spherical and quite large. The delicate blues and coppers (Lycaenidae) often have a deeply etched honeycomb or scalloped pattern. Eggs are laid in batches sometimes, such as those of the Black-veined White (*Aporia crataegi*). The Milkweed or Monarch (*Danaus plexippus*) lays its eggs singly or in small clusters like most of the swallowtails (*Papilio*). *Malacasoma* moths lay theirs clustered in tight rings round a twig and cover them with a tough glue which protects them from their winter ordeal. Some species simply scatter their eggs loose in the undergrowth as they fly along but each kind has its own particular habit from which it seldom deviates. The females instinctively know where is the best place to lay, ready for the hatching of the larvae. The Silver-washed Fritillary (*Argynnis paphia*) lays her eggs, not on the leaves of violet which feed the larvae but on tree bark nearby because the larvae hibernate here before descending to the leaves in spring. The female of other species singles out the exact foodplant to lay her eggs on, using a detection system that is quite unknown to man. Many eggs change colour after laying, indicating that they are fertile, and again just before the caterpillar hatches.

Left
Until the early part of this century the Black-veined White (*Aporia crataegi*) was widely distributed in Britain and even known as a pest of fruit orchards. Now it has disappeared and is almost certainly extinct, though commonly found in most other parts of Europe.

The newly hatched caterpillar usually takes its first meal from the eggshell and many will not survive unless they do this. It is an important point to remember when breeding butterflies and moths. The caterpillar then starts feeding from the leaves of its foodplant which might be nearby or right under its feet. It is a tiny creature compared with the size it will eventually become and it will increase its weight by as much as 1 000 times. Its very purpose is to eat and grow and some caterpillars hardly take a rest. The skin is only capable of stretching within certain limits so the caterpillar sheds its skin several times during its growth and a new, larger skin appears beneath the old one which splits down the middle and is shrugged off at the tail end. Most species change their skins about four times though as many as forty changes have been known.

Right
The Purple Emperor (*Apatura iris*) is the grandest of the European woodland butterflies. The purple colouring is seen usually only on one side at a time, according to the angle of vision, but, occasionally, the full splendour is caught with the rich colouring seen across all four wings. This nymphalid shares with *Charaxes*, *Prepona* and *Agrias* their love of carrion, rotting fruit, and so on.

Left
The pupa of the Orange Tip (*Anthocharis cardamines*) is shaped very unusually, looking like a thorn or seedhead, so that it escapes attention while it rests the whole winter through. The two colour forms are not necessarily related to the colour of their background but usually the green form is found in places where there is greater light intensity.

Particularly in temperate regions, there is a dormant stage in the life history which is hibernation (sometimes in the tropics there is aestivation during the dry season). This can be at any one of the four stages, according to the species, but very often it is in the caterpillar (larval) stage. Rather few hibernate as fully grown larvae (perhaps if they did they would fall victim to predators more easily), some hatch from the egg and hibernate without feeding, and others start their growth, hibernate and complete the cycle in the spring.

The caterpillars of butterflies and moths are as varied and interesting as the adult insects themselves. There is no limit to the humps, spines, iridescent blobs, whips and brushes that nature uses to decorate and even conceal this stage of the life cycle. Larvae of the prominent moths (Notodontidae) are often very angular in shape with markings that accentuate the unusual form making it unrecognizable even though very conspicuous. An example is the larva of the Puss Moth (*Cerura vinula*) which has twin whips at the tail that eject long pink flagellae in an unexpected and frightening manner if a curious predator comes too close. Another is the Lobster Moth (*Stauropus fagi*) so called because of its long forelegs (not found on other larvae) and bulbous tail portion of the body, as well as its pinkish colour.

The caterpillars of Lycaenidae (the blues, coppers and hairstreaks) have a characteristic shape that is sometimes described as slug-like, being tapered at each end and humped, though this is where the similarity ends because they are usually beautifully coloured, often with markings that make them practically invisible on their foodplant. The Chalkhill Blue (*Lysandra coridon*) larva even becomes yellow and green to blend in with the bright yellow flowers of Horseshoe Vetch among which it feeds in summer. Harsh spines protect many caterpillars, especially

Top right
Typical of several of the swallowtail caterpillars, this rarity from Corsica, *Papilio hospiton*, is feeding, not on its native foodplant, but on the leaves of cultivated parsnip, which it has not previously been known to eat.

Right
Limenitis dudu from Asia is well concealed as a pupa by the extraordinary shape and uneven markings which make it appear to be a shrivelled leaf. The butterfly sits with open wings for long periods, basking in the sun, like many species of *Apatura*.

Far right
Puss Moth (*Cerura vinula*) caterpillars seem to be very bright and prominent when seen out of context, but in the wild they rest in postures that imitate the way poplar leaves turn black at the edges. The caterpillar has twin tails that eject long red flagellae if it is disturbed and it can also spit formic acid with a remarkably good aim.

Left
The defensive attitude of the caterpillar of *Papilio demodocus* is particularly effective when it rears up its head in this manner. The eye markings are purely to frighten. Behind the head a forked organ, the osmaterium, is ejected and the air is filled with a strong smell of rotting fruit.

Right
Some pupae do not have the legs, antennae and other parts fully enclosed inside the pupal capsule. This pupa of the Asian Banana Skipper (*Erionota thrax*) is a particular curiosity because of its separate and very long proboscis which extends even beyond the tip of its abdomen.

those of the nymphalid butterflies, and some live gregariously –a greater protection against birds that might be more put off by a cluster of prickly morsels than a single one. The spines of the South American *Automeris* moths are highly decorative but they sting like a nettle. One related species causes a fatal disease and bleeding which cannot clot; this Venezuelan horror is the only truly harmful lepidopteron known, though it is always wise to avoid handling hairy caterpillars as they can cause skin irritation.

Silk is spun by most caterpillars, primarily as a foothold on the surface that they rest on or continually use, and, in the case of many moth species, silk is used to make the cocoon. A pad of silk is spun as an anchor when a caterpillar changes its skin and I have even seen a caterpillar 'motorway' of silk constructed the entire length of the trunk of a tree, along the ground and up another tree where a group of processionary larvae were living. It is fascinating to watch a caterpillar weaving a cocoon around itself, drawing out a thread which is a fast-setting liquid, from just under its head.

Each time the caterpillar changes its skin there is another skin which has developed inside the old one. It usually has the same colouring and shape as the old larval skin and the only difference is that it is larger. When the final change comes and the caterpillar is ready to form a pupa, the process is much the same but, instead of there being another larval skin beneath, the new one is in the form of the pupa. It is still soft, usually not the colour of the final pupa and still rather larval in shape, but within an hour or so the pupal skin hardens, forming a shape which is quite unlike that of the caterpillar, and usually taking on its own colour or pattern. The pupa is vulnerable since it cannot see or move. It is usually attached or enclosed either inside a cocoon or other shelter and has to be fixed in a concealed place. The pupae of many butterflies, especially Papilionidae, Pieridae and Lycaenidae, are usually attached by the cremaster at the tail end of the pupa, to a pad of silk. The caterpillar spins a girdle of silk around its thorax and what will become the pupal wing-cases, to hold the pupa usually in an upright position. The typical position for the pupa of a nymphalid is hanging head downwards, supported only by the cremasteral hooks which catch in a firmly secured silk pad. An interesting situation arises when the caterpillar sheds its larval skin having firstly anchored the tail. The skin is wriggled down to the tail end and has to be disposed of–but remember it is hanging from the tail and has nothing else to stop it from falling! The answer to the problem is sheer speed of movement; the skin has to be ejected in the split second it takes to release its grip, flick off the skin and catch hold again before it can fall and this the pupa carries out as if it has been doing so all its life.

Butterfly pupae that are suspended or supported by a girdle, even when in a place concealed from the eyes of predators, are very exposed to the elements. They dehydrate less easily than pupae that are protected by soil or a cocoon, and they are usually harder. A swallowtail pupa that is fixed to a reed on the Norfolk Broads in England has to withstand frost, snow, ice, wind, sunshine and even total immersion in water. By contrast, the pupa of the Lappet Moth (*Gastropacha quercifolia*) is so soft that it can be damaged almost by the touch of a finger. It is enclosed in a cocoon and has only to withstand the best of summer weather over just a few weeks. The Puss Moth finds itself a novel protection for the winter. The larva chews up the bark that it scrapes out to make a hollow on a tree trunk, and mixes it with silk to make a concrete-hard cocoon which is so like the bark that it is barely visible. Hardly any butterfly pupae are formed underground but the larvae of many moth species burrow, either just below the surface, or to a depth of several centimetres. Here they make a little tomb-like hollow or even spin a light silk cocoon.

Many pupae and cocoons are beautiful to look at or incredibly well camouflaged. The gold flecks found on some of the danaids and nymphalids are like jewels. They look like dew drops or even chinks of light seen through a dead leaf, giving convincing camouflage. It is this gold which gives us the name chrysalis, derived from the Greek word for gold, *chrysos*. Tropical *Euploea* pupae carry this even further by being totally silvered like a mirror and so reflect whatever background they are among.

The pupa of some of the aristolochia-feeding swallowtails, *Papilio alcinous* for instance, is curiously gnarled and twisted. It is rather reminiscent of a chewed toffee and not at all like a chrysalis! The cocoon of the Japanese silkmoth (*Dictyoploca*) is a beautifully formed net, as uniform as a fisherman's, through which the pupa is clearly visible. *Rhodinia fugax* spins a cocoon of green silk, shaped like a pitcher plant and open at the top. The open cocoon does not fill with rainwater because the caterpillar spins a clear round drainage hole in the bottom, and this prevents the pupa from drowning!

The change from the larva to the pupa provides a resting stage in which the internal organs can develop and reconstitute themselves. Although the inside of a caterpillar is largely taken up with the well-developed digestive system there are also embryo features of the adult insect. The pupa is an immobile capsule that has no legs and it cannot feed, but inside it the nerve system enlarges and so do the imaginal buds (the term given to the groups of cells which multiply and build themselves into the parts of the adult, such as the wings, legs and antennae). So that, although the pupa looks quite different to us, the inside is still rather like that of the caterpillar until the point when the adult has formed properly inside, a little like the formation of an embryo inside the egg.

Left
The pupae of *Euploea* species have a reflective surface which has the effect of taking on the colouring of its background as mirrors would. This *E. midamas* pupa is particularly silver, while others are metallic bronze, blue or green. All these colours were shown by a single brood of pupae from Hong Kong, from which this pupa came.

Top right
Maculinea arion, the Large Blue, has a curious life-history discovered in 1915 by Captain E. B. Purefoy in England. The caterpillar wanders off its foodplant and is taken into the nest of a red ant where it is nurtured on the ant larvae, giving in return a sweet honeydew from a gland on the back. This unique photograph shows the caterpillar in a walnut-shell nest used experimentally to rear the species in captivity—a feat achieved only once since the original findings in 1915.

Right
The Large Blue butterfly, shortly after emergence from the ant nest. The species is fighting for survival in Britain, helped by attempts to breed it and special ecological work by the Nature Conservancy.

Adult butterflies and moths

Above
A close-up portrait of the head of the Elephant Hawkmoth (*Dielephila elpenor*) shows the large composite eyes (each one having thousands of facets forming a single image), and the coiled proboscis which it extends deep into the centre of flowers to take nectar, even in flight, like a hummingbird.

In English-speaking countries there is more accent on distinguishing between butterflies and moths than is, perhaps, justifiable. In certain languages there is barely any differentiation. Scientifically, certain families of Lepidoptera have been placed under the heading Rhopalocera (butterflies) and others under Heterocera (moths). The selection is to some extent arbitrary but the characters which designate the separation of butterflies from moths are about as consistent as anything can be in nature. There are generally exceptions to such rules but the one with the fewest exceptions is that butterflies can be recognized by their antennae which have a rounded or clubbed end, whereas those of moths may end in one of several other ways, possibly finely pointed or even pectinate (comb-like). In principle, moths settle with their wings held back over the body while butterflies either hold them upright or spread out flat either side. A glance at the Lepidoptera families quickly explodes these broad principles and shows up many exceptions but you can be reasonably guided by using the antennae criterion.

The body of a butterfly and other insects is constructed quite differently to that of the human frame or a bird's. Instead of an internal skeleton an insect has an exoskeleton, which is a series of plates (resembling a suit of armour) joined by softer membranes. The exoskeleton is usually flexible and is made of chitin which is not unlike the material that our finger nails are made of.

There are three divisions of the body, the head, thorax and abdomen. The head is largely taken up by the compound eyes which are situated on either side of it. These eyes are quite unlike our own, being made up of thousands of facets. The compound eye is very sensitive to light but does not have very accurate vision. It has a very wide angle of vision and records movement, warning the butterfly of danger. Between the compound eyes there are usually three minute single eyes (ocelli) and the two antennae which are the most important sensory organs, passing information directly to the brain. The sense of taste, curiously enough, comes from the feet and hearing from a tympanum on the thorax. Food is taken in through the proboscis which is coiled up like a watch spring under the head when not in use. Two separate parts are fused together to form a tube which can probe deeply into a flower for nectar. Certain moths have no proboscis or only a rudimentary one and they can live without feeding. There are no chewing mouthparts in butterflies or moths.

There are no veins, arteries or capillaries in the circulatory system of an insect. The blood is restricted only by the limits of the insect's body cavity, the haemocoel, in which lie the various organs of the body. The tissues are thus bathed in blood rather than supplied by a system of vessels. Therefore, an insect, particularly a soft caterpillar, bleeds to death if it is punctured. The blood is pumped round the body by the heart, a long, many chambered tube which lies along the back. The blood enters at the hind end and along the sides (valves prevent backflow) and is pumped out at the head end. Breathing is by a series of tubes which branch from breathing holes (spiracles) situated along the sides of the abdomen. The tubes divide and become finer as they go deeper into the body tissue and air is passed through the system of trachea and tracheoles by capillary action. The nervous system consists of a chain of nerve centres, ganglia, connected by a ventral nerve cord which leads from the head along the underside to the tail. The 'brain' is a simple fusion of the ganglia corresponding to the head segments and does not exert the same degree of control over the creature that the brain of a mammal does. The digestive system of the adult is degenerated from that of the caterpillar as it is now less important, but the nerve and reproductive systems are fully developed.

Below
Skipper butterflies will sometimes assume the resting wing position typical of moths. *Erynnis tages* rests like this by night but by day it remembers to be a butterfly and alights with its wings outspread!

Wings and flight

The pattern on the wings can often be seen through the wing cases of the pupa just before the adult emerges. On emergence the wings are as tiny as they look inside the pupa and the markings are seen in miniature. The adult finds a place where it can hang, wings downwards, without obstruction of the wings which now begin to expand so fast that you can sometimes actually see them growing. Air is taken in and pressure built up inside the body, forcing blood into the nervures of the wings until the upper and lower membranes are brought together and almost join. An hour or more passes before the wings are dry and fully able to bear the weight of the insect in flight. For efficiency in flight the two wings on each side have to be linked. In moths this is achieved by a bristle and catch mechanism and butterflies have a lobe on the hindwing which overlaps the forewing and grips it in flight. The species with smaller wings in relation to the size of the body have a fast wing beat while those with a very large wing area glide with a slow and graceful flap of the wings. The speed of flight also varies with the build of the butterfly or moth. A Painted Lady (*Vanessa cardui*) averages around 24 kilometres per hour (15 miles per hour), the Large White (*Pieris brassicae*) seldom exceeds 10 kilometres per hour (6 miles per hour) unless frightened when it can double this speed. The faster hawk-moths are capable of about 56 kilometres per hour (35 miles per hour) and appear to fly even faster than that as they dart about in search of nectar or on their purposeful migratory flights.

The emergence from the pupa and drying of the adult wings are always fascinating to watch. On emergence the wings are tiny in comparison with the body, as seen in this Broad-bordered Bee Hawk (*Hemaris fuciformis*). The insect finds a climbing point and eventually rests with wings hanging downwards, well clear of the ground or obstacles. The wings are held slightly apart and they can often be seen visibly growing. Within only about twenty minutes the wings can be fully expanded but another hour or so is needed for them to harden and the moth or butterfly can then assume its more normal wing resting position. The Bee Hawks are extraordinary in that the light scaling that covers the wings on emergence is shaken off during the first flight, so they become quite transparent.

Left
Milkweed Butterflies or Monarchs (*Danaus plexippus*) are exceptionally strong and well known for their migratory feats north-south across North America and even right across the Atlantic Ocean to Europe. The larvae feed on *Asclepias* which is not hardy, and because of this the Milkweed has been unable to establish itself elsewhere in Europe as it has done in the Canaries.

Below left
This delicate Long-tailed Blue (*Lampides boeticus*) is a habitual migrant, even able to fly the English Channel to the coast of southern England where it sometimes breeds. *L. boeticus* is also found in Asia, where this one came from, and from Africa. The larvae feed on vetches and will even thrive on podded garden peas!

Migration

Butterflies are capable of flying long distances on migratory journeys. Comparatively few species of Lepidoptera migrate and, unlike birds, they live too short a time to fly out for the winter and back again in the spring. One of the most famous of all migrants is *Danaus plexippus*, known as the Monarch in America, the Milkweed in Britain and in Australia as the Wanderer. While its purposeful flight could hardly be described as wandering it is a fact that this butterfly has succeeded in flying from America to Australia where it is now firmly established. It also now lives in the Canary Isles and it flies the Atlantic Ocean from North America to Britain. Its regular migrations from the northern states of America to those of the south are witnessed each year and the butterfly trees where they pass the winter hanging in thousands are now a tourist attraction. Several British species cannot survive the winter and arrive each summer in migrations from as far away as North Africa and the Mediterranean. These include the Red Admiral (*Vanessa atalanta*), the Clouded Yellow (*Colias crocea*) and the most universal migrant butterfly of all, the Painted Lady. In the tropics *Catopsilia* species and other Pieridae are seen in very dense flights, by which they are

able to extend their range and move from areas which are suffering from drought. In the Pyrenees there are particular passes through which butterflies, moths, dragonflies and other insects regularly migrate on their journey northwards in the spring and (their progeny) south again in the autumn. When using a mercury vapour light to attract moths one night on the English coast, I became involved in a migration of Silver Y Moths (*Plusia gamma*) and a quantity of something like 6 000 descended upon me and my lamp, which put an end to a quiet night's study and caused quite a bit of confusion!

Above
Occasionally butterflies find their way into foreign parts by hitching a lift in cargo. This *Opsiphanes* arrived in a wholesale fruiterer in Somerset, England among crates of bananas which had come from Brazil. The caterpillars feed on banana leaves and one had probably pupated among the fruit.

Right
Through the Pyrenean passes Lepidoptera flock with dragonflies and other insects each summer. One of the hawkmoths which migrates this way is this Humming-bird Hawk (*Macroglossum stellatarum*) which hovers and darts like its namesake and is hardly ever seen like this, resting on a flower.

Below
The Painted Lady (*Vanessa cardui*) is most noted of all the Lepidoptera for its migrations. Although sometimes seen in vast numbers, more commonly they spread out and come to gardens in small numbers.

Right
Distinguishing between the sexes of many species is difficult but some obliging Lepidoptera, as this Cherry Moth (*Callosamia promethea*), have a colour difference so great that they could be taken for two different species. The male here is black and the maroon moth is the female.

Light trapping

A number of theories have been put forward as to why moths are attracted to light. Of course it is not only moths but practically any night-flying insect and certain butterflies that are crepuscular which come to light. The ultra violet in a mercury vapour lamp attracts insects more strongly than tungsten light and specially made traps have been designed to enable a collector to leave his light unattended at night. Moth trapping is an excellent way of learning to identify all the species. It is not necessary to kill the moths (and an anaesthetic chamber in the trap should *never* be used) but, after those that are wanted for study have been taken, and the rest identified, the catch can be released into long grass the following evening. Unless the moths can quickly find cover when released they will be eaten by birds and this is why long grass is suggested.

Distinguishing the sexes

Distinguishing between male and female is not always easy, especially with butterflies, but by careful examination of the abdomen and genitalia it is possible to see a difference. The female abdomen is generally fatter and, at the tail, more rounded than the male, which has a more pointed tip. The male has claspers–triangular plates with which he grips the female when mating. Females sometimes have more rounded wings but this difference is usually very subtle. Some males, for instance those of skippers, euploeas and fritillaries, have patches of scent scales (androconia) which are prominent and give off a faint perfume to attract the female in courtship. A good many species are sexually dimorphic: the Brimstone, for example, has a bright yellow male and a female that is a pale whitish green. The females of morphos are usually brown instead of the famous iridescent blue of the male.

Scent attraction and courtship

The females of many moth species are able to attract males by the emission of a scent (which we cannot detect) from glands at the tail. I have had a female Emperor Moth (*Saturnia pavonia*) attract about forty males when I took her to a piece of heathland which was a known locality for the moth. The female vapourer moths (*Orgyia*) are not winged; they remain on the cocoon from which they emerged and attract the male in this way. The wingless female hardly

moves, in fact, and the eggs are laid all over the cocoon soon after mating. In courtship it is often the male that uses scent and some of these scents can be detected by human noses. The Danaidae males have scent brushes which they unfold and use to spread the scent all around the female, also making a most attractive display. The androconia, one of the features which enable us to identify the males of a species, are sometimes hidden in the abdominal flap of the hindwing, sometimes they are in a bump or pouch on the upper surface of the wing or they may be plainly seen as a broad patch or series of bands across the wings. There is often a nuptial flight, particularly with butterflies, and some are almost impossible to breed in captivity because of the need for this flight which sometimes takes them high up in to the tree tops. Other species are content with a simple courtship flight in quite a restricted space and I have had the opportunity to watch this closely in a simulated jungle where I breed South American *Heliconius* butterflies and some of the Papilionidae: it is charming to watch and something that can be set up without too much difficulty in a greenhouse. Butterflies stay paired, tail to tail, usually for an hour or two. Many moths remain paired for at least twenty-four hours, from dusk to dusk.

Feeding

One of the most important requirements of butterflies and moths is moisture rather than food; they dehydrate very quickly. In dry weather butterflies live much shorter lives and they can be kept alive on water alone. The commonest food for butterflies is nectar and they have distinct preferences for certain flowers which provide the most or best. Moths have similar feeding habits, though some get by on water alone and have no proper mouthparts for taking food. Moths are drawn to sugary secretions and in fact may be attracted by spreading the right sweet solution on tree trunks, a method known as treacling. Butterflies, particularly Nymphalidae, will be drawn to exuding sap and some are found drinking from carrion, dung and rotting fruit. The diet of butterflies includes protein obtained from the pollen of the flowers they visit; some, and I have watched this in great detail in Brazilian *Heliconius*, secrete a liquid on to a pile of collected pollen, dissolve it and then drink the enriched liquid. Certain salts are beneficial to butterflies and in the tropics it is a common sight at river banks or even patches of urine, to see butterflies densely flocking together probing for moisture and salts.

Below
One of the sights to look for in the tropics is a flock of butterflies congregating at a damp mud patch. This group in Africa includes the swallowtail *Papilio demodocus* and some Pieridae which are mostly *Belenois* species. The butterflies take up various salts and minerals as well as moisture and will often be attracted to urine.

Coloration

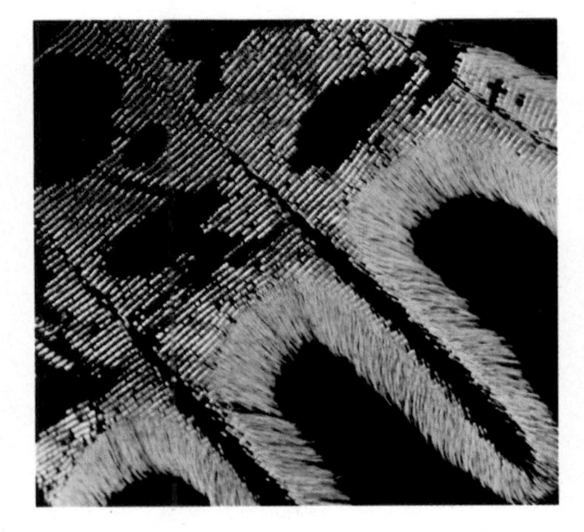

Above
The iridescent colours on the wing of this day-flying moth (*Urania ripheus*) from Madagascar are not pigmental, but are caused by the scale structure refracting the light at different angles. The uniform white fringe to the wings is made up of thousands of pure white pigmented scales.

The patterns and wing markings of Lepidoptera are made up of a mass of tiny coloured scales which overlap each other, rather like the tiles on a roof. The colouring in the scales is caused either by a pigment contained in the scales themselves or by the structural character of the scale which refracts light to give off an iridescent colour, though it contains no pigment.

Pigmental colours are either formed from a chemical substance within the insect itself or they can be derived from the foodplant of the caterpillar. The yellow colouring of the Pieridae is produced by substances known as pterines which are derived from excretory uric acid. The red and orange pigments in the Vanessidi are not related to uric acid and almost certainly come from the larval foodplants. Interesting experiments can be carried out with the extraction and changing of pigments. The red pigments just referred to are affected by exposure to the oxygen in the atmosphere. A freshly emerged Painted Lady is much brighter than one a week or so old. However, if a faded specimen is exposed to chlorine the colour is restored or made even brighter than the original. Flavones, another important group of pigments, come from the foodplant and are responsible for the colours in flowers ranging from ivory to yellow. Blacks are derived from melanin which produces freckles in human beings and the black pigment in piebald animals. Green pigments are very rare in butterflies and not very common in moths. Green colouring is more often caused by refraction and blue is always due to refraction.

The orange tips (*Anthocharis*) appear to have a green dappled marking on the underside of the hindwings (an excellent camouflage) but this is an optical illusion. The pattern is, in fact, made up of an intermingling of black and yellow scales which are clearly seen as soon as enlarged under a good lens. Some colours appear to change according to the angle of vision and an interesting example of this is seen on the hindwings of the birdwing *Troides magellanus* from the Philippines and Taiwan. Seen from above the colour is golden yellow but if viewed from the tail across the plane of the wing (at eye level) towards the head, the colour is a luminescent electric blue. A South American *Pierella* has forewing patches which change to at least four different iridescent colours according to the angle of vision.

Camouflage

One of the most important reasons for the colouring of butterflies and moths is to camouflage them and the Lepidoptera have some of the most fascinating examples of camouflage in the animal kingdom. Quite a number bear a strong resemblance to leaves and perhaps the most effective of these is the Leaf Butterfly (*Kallima*

Left
This Indian Leaf Butterfly (*Kallima inachus*) seen resting on the ground among other leaves would escape the attention of the most keen-eyed bird. This must be one of the world's best camouflaged creatures.

Below
Winter for this Brimstone butterfly (*Gonepteryx rhamni*) is spent hibernating in foliage, usually in ivy. It takes little imagination, therefore, to see how the butterfly gets its unusual shape and colouring.

Right
Orange Tips (*Anthocharis cardamines*) have a mottled green patterning on the underside and this picture shows how well this conceals them when they rest on the flowers of their foodplant, Garlic Mustard, or the similar bunches of white florets here.

Above
The beautiful colouring of the Orange Tip shows when the butterfly opens its wings. Only the male has tips of orange.

Above
Peppered Moths (*Biston betularia*) produce two forms which freely interbreed (the two illustrated here are pairing). The dark form prevails in polluted areas and in cleaner countryside it is the white form that is most commonly found. Natural selection undoubtedly brings this about since one form is bound to stand out more than the other in either type of habitat.

inachus) from Asia. The wings are the exact shape of leaves and the marks on the underside closely resemble the midrib and veins. No two butterflies are exactly the same shade of colour. Some are marked with spots like leaf decay. The butterfly alights with head downwards and its tail touching the stem so that it appears to be growing out of it. The Brimstone (*Gonepteryx rhamni*) hibernates among evergreens, usually bushy growths of ivy, and the colour and shape of the Brimstone's wings are a remarkable imitation of the ivy leaves among which it settles for the winter. The Chinese Character Moth (*Cilix glaucata*) has a curious resting attitude and the exact colouring of a bird dropping. The Elephant Hawkmoth (*Dielephila elpenor*) is a beautiful pinkish maroon – very conspicuous when out of place, but when it is among the flowers of its foodplant, Rosebay Willowherb, it exactly matches the magenta flowers. There are even moths that are so like the bark or lichens where they hide during the daytime that they simply melt into them and are very hard to see even when you know they are there. The Peppered Moth (*Biston betularia*) settles often in an exposed place but matches its background, only it goes one better than most. It has developed a black form which now appears in the black industrial areas while the paler, peppered form is still found in unpolluted areas.

Eggs tend to rely on being concealed rather than camouflaged but the females of some moths are equipped with tufts of hair on the abdomen expressly for the maternal duty of camouflaging the newly laid egg batch. When eggs are camouflaged they usually resemble a bunch of seeds.

The camouflage of caterpillars cannot fail to impress the naturalist. The stick caterpillars are some of the most fascinating; (these are not to be confused with stick insects which are in a different order). They are capable of taking on the exact shade of colour of the twigs that they resemble and caterpillars taken from the same brood but reared on different bushes will be different colours. These caterpillars, which are all in the family Geometridae, have evolved bud-like projections, scars and other bark-like markings so that they are most difficult to detect, especially as they rest at an angle to the twig, head outwards, forming a natural looking V-shape. One of the best ways to find stick caterpillars is to shake branches or tap them with a stick over an umbrella or sheet

Right
It is interesting to see how this Peppered Moth caterpillar has taken on the exact colouring and texture of the branch on which it is living. It is even possible for them to be purple or bright green if that is the colour of their foodplant.

Below
Effective camouflage is demonstrated by this caterpillar of the Orange Tip (*Anthocharis cardamines*) which is counter shaded so that it blends with the stalks that it lives on.

Above
Chalkhill Blue (*Lysandra coridon*) larvae feed among the Horseshoe Vetch of downland and when the yellow flowers are out they take on a colouring of green and yellow which allows them to feed quite openly without being recognized.

and you may well dislodge some, along with a host of other interesting creatures. Woodland, scrub and forest are the best types of locality.

As with certain moths there are caterpillars which might well be mistaken for a bird dropping. Larvae of comma butterflies (*Polygonia*) are protected in this way with markings in black or brown and white. It is also a characteristic of a great many of the early instars (stages between the skin changes) of the larvae of the swallowtails (Papilionidae). Those of *Papilio demoleus* from Asia are positively glistening as if very freshly deposited, but later the caterpillars lose this camouflage protection and become conspicuously coloured with either frightening eyes or warning colours. The caterpillars of the lappet moths (Lasiocampidae) have outgrowths along their body which enable them to wrap themselves around the twig of their foodplant and, as they are already rather well camouflaged, all you see is a slight thickening of the branch unless you happen to touch one when it will rear its head and show bright orange patches. Both the European pine hawks (*Hyloicus*) and an American sphinx (*Lapara bombycoices*) have larvae that feed among pine needles and are so striped along the length of the body that they are most difficult to see. When the pine hawk is in its final instar it feeds on the thicker branches and its striped skin is cast in favour of a mottled brown pattern which exactly matches its new feeding place.

Another aspect of camouflage that is very effective, is that of counter shading. The caterpillar of the Eyed Hawk (*Smerinthus*) is typically marked with oblique stripes along the sides of the body and the characteristic hawkmoth tail. However, the stripes are coloured to resemble the veins of the leaves of its foodplant and the colour is much darker on the ventral surface than on the dorsal. Therefore, while it rests in a natural pose it resembles its foodplant but if it were to rest upside down it would immediately stand out and its whole shape become prominent. The caterpillar of the Squeaking Silkmoth (*Rhodinia fugax*) from Japan shows this even more by having its top half a very pale lime green and a distinct dividing line where it becomes dark green beneath.

While a great many pupae do not rely on camouflage as much as concealment for their protection there are some striking examples in those which tend to be more exposed. The Purple Emperor (*Apatura*) forms a pupa among the sallow leaves that it feeds on and it is the exact pale, rather whitish green that is the colour of the back of the leaves, with the same prominent vein-like markings. Once again, the bird-dropping theme is used by the *Theclas* (Lycaenidae). Some of the swallowtails (Papilionidae) have pupae that are not only patterned to look like wood but have an irregular shape that looks like a broken-off twig. Some even have bright green markings which look like natural algae or lichen. *Heliconius*, *Polygonia* and other Nymphalidae have pupae that are like withered leaves, often with strange projections or even shining silver and gold marks to break up the shape.

Below
Heath Fritillary larvae (*Melicta athalia*) are remarkably like the seed heads of plantain on which they feed. Their dark colour would make them conspicuous on the leaves alone.

Above
Superb camouflage is shown by the pupa of the Purple Emperor (*Apatura iris*) which is formed among the leaves of the food-plant, Sallow. It is exactly the colour of the underside of the leaf and even shows midrib and veins.

Warning colours

Warning colours are displayed by Lepidoptera that are distasteful to predators and some of these are distinctly poisonous. The toxins are usually derived from poisons that exist in the larval foodplant, and to advertise that they are unpleasant to eat they are distinctively coloured and patterned in contrasting shades of red, yellow and orange with black. The Milkweed Butterfly caterpillars are impressively ringed in yellow and black and birds generally ignore them. If an inexperienced young bird does try one it will produce nausea and illness and the bird will not try eating another caterpillar which has that pattern. Related Danaidae, including the *Euploea*, which are coloured in deep violet and black, are also protected by being poisonous in all the stages of the life cycle. The black and orange ringed caterpillars of the Cinnabar Moth (*Callimorpha jacobaeae*), feeding on ragwort and groundsel, are familiar sight in summer and these too are protected by warning colours which, it will be noticed, are the same as those displayed by wasps and other stinging Hymenoptera.

Mimicry

It is a natural progression of these warning colours that certain species take advantage of this protection and wear the uniform of a poisonous species when, in fact, they are not. In North America the nymphalid *Limenitis archippus* bears exactly the colouring and marking of the Milkweed and, although it is noticeably smaller, it succeeds in fooling birds and escapes their attention even though it is perfectly good to eat! This is known as Batesian mimicry, after the naturalist Bates who evolved the theory. Bates found that there were numerous examples of such deception among the Lepidoptera, particularly in the tropics. In Africa the nymphalid *Hypolimnas misippus* mimics a model that is closely related to the Milkweed but it is only the female that has this protection–she must live longer to lay the eggs. *H. misippus* is very artful here, because in certain parts of Africa its model, *Danaus chrysippus*, has a different form which lacks the typical white bands on the forewings, so in these regions, *H. misippus* follows suit with a matching form with no white bands. In Malaysia there are black and violet papilios which mimic the poisonous *Euploea* species and a satyrid (*Elymnias*) which produces

Left
Warning colours of yellow and black in striking contrast make predators shy of tasting the caterpillars of the Milkweed Butterfly (*Danaus plexippus*). They are capable of making a bird very ill, and the poisons in their body which come from the foodplant can even kill a predator.

Below
The Malay Lacewing (*Cethosia hypsea*) mimics the poisonous *Danaus chrysippus* particularly effectively when in flight. The spiny caterpillars are ringed in red and yellow and live gregariously, feeding on passionflowers.

several completely different colour forms which mimic bright red and yellow *Delias*, brown and white mimics of *Danaus chrysippus*, dappled green forms to imitate certain danaids and purple forms to match *Euploea*. Certain moths convincingly mimic wasps with bulbous yellow and black bodies and transparent wings. Fritz Müller evolved a variation on Bates' theme which is known as Müllerian mimicry. He noticed that some distasteful species mimic others and the advantage is not so obvious as a palatable species mimicking one that is distasteful. The fact is, however, that if predators have to learn to recognize only one combination of colour and pattern instead of several, many fewer butterflies will be lost before that pattern is recognized and avoided. Thus, in South America, it may be possible to find a dozen butterflies flying round a particular bush, all quite different species but all wearing the same uniform of red and black bands. Another mimetic colour combination used widely in South America by both butterflies and day-flying moths, is orange and black. A collection of scores of similarly patterned orange and black species, apparently mostly the same, might in fact not have a duplicate among it and the study of these examples is truly fascinating. Mimicry becomes really complicated with the *Heliconius* since their shape is basically the same but *H. erato* in some regions takes on the quite different colours of *H. melpomone* and vice versa, so how are we to tell them apart? The same happens with many of the other helicons and the only way entomologists can be sure of identification is by examination of the genitalia or by breeding since the larvae or pupae may be quite distinct.

Startling colours

Most of the Saturniidae or silkmoths have prominent, round, eye-like markings which make a rather frightening spectacle to a prospective predator. A brassolid, the Owl Butterfly (*Caligo*), has similar markings which are very effective when it suddenly flips its wings open and closed in the face of a predator. *Automeris* have the eyes on the hindwings which are concealed until they are disturbed when they are flashed open, displaying brightly coloured target-like markings. Such flash colouring is adopted by the Peacock Butterfly (*Inachis io*) whose brightly coloured upperside is concealed by the black, camouflage pattern of the underside. When disturbed the Peacock opens its wings very suddenly and makes a loud rustling sound. The catocalas or underwings and certain tiger moths rely on this startling effect of flash colouring when they open their wings very suddenly, often taking flight before the investigating creature comes to its senses. *Morpho* butterflies also have a very effective brilliant blue flash colouring. They have the additional protection that they are most difficult to trace as they fly off; with the downward

Below
Young Lobster Moth caterpillars resemble red ants not only in their appearance, but also in their habit of clustering like ants on a branch in search of honeydew. They even move their false legs constantly like the movement of ants' antennae.

Below
The Lobster Moth (*Stauropus fagi*) caterpillar is protected by its peculiar, elongated forelegs which it shoots out in a frenzied manner if a predator comes too close. The extraordinary shape is a disguise used by many others in this family, the prominents (Notodontidae).

Top
The model which sets the style for more mimics than any other is *Danaus chrysippus*. It is found throughout Africa, Asia and Australia. Closely related to the Milkweed, the caterpillars feed also on milkweeds (Asclepiadaceae). A number of different geographical forms exist.

Above
The clearwing moths (Sesiidae) are very convincing mimics of stinging Hymenoptera. Some are very slender indeed, while others, which include the hornet clearwings, are robust and might well be taken for hornets. This orange moth is perfectly harmless, but would you not hesitate to pick it up?

wingbeat they are blue, and hard to see against blue sky, and then they are brown with the upward beat, so that one soon loses track of them. The Ghost Moth (*Hepialus humuli*) has this trick too. The male is shining white above and black underneath. Its name comes from the way it seems to appear and disappear again when hovering over grasses at night.

Diversionary markings

The rather smaller eye-spots towards the tail of some of the satyrid butterflies, or in the centres of the wings have a different purpose. Rather than frightening, they are so gleaming and eye-like that a bird will often peck at them causing relatively little damage and enabling the butterfly to escape. Many *Papilio* species have an eye at the tail which is rather less convincing but undoubtedly helps to confuse an assailant. The markings at the forewing tips of *Attacus* moths are so incredibly reminiscent of a snake's head, with eye and mouth clearly defined, that these must surely be rather off-putting, as well as the fact that these atlas moths are twice the size of the average bird. Certain hairstreak butterflies (Lycaenidae) have tails that are sufficiently antennae-like for a bird to mistake them when the butterfly rests, head downwards, shuffling its wings so that its 'antennae' convincingly move like real ones. Little damage results if only the tails are pecked off!

Disruptive patterning

I seldom walk in the country without having my eyes open for whatever interesting plant or creature may present itself. On walks to school as a child I would find all sorts of caterpillars or moths at rest and my eye became trained to notice even camouflaged creatures. A bird is probably even more attuned to this if it relies on finding insects as food, but if the conventional shape of a creature is sufficiently broken by some prominent mark or bar slashed across it, there is every chance that it will not be seen. The unusually shaped larvae of the prominent moths (Notodontidae) have already been mentioned and the same principle is applied to the pupae of several tropical butterflies, and especially to geometrid moths, such as the Bloodvein Moth (*Calothysanis amata*), which rest by day with their wings outspread. *Limenitis* pupae have a very disruptive pattern and reflective chinks to deceive the eye.

Variation

Some butterflies have different colour forms for the wet and dry seasons or for winter and summer. The African *Colotis* butterflies have darker markings in the wet season and *Precis octavia* has two forms that are so unlike each other that they could be taken for different species. The European Map Butterfly (*Araschnia levana*) is fritillary-like in spring, being chequered in black and orange, but the summer form is black with a broad white band rather like a White Admiral. Geographical variation is shown in a number of butterflies. It is often possible

Above left
The *Heliconius* or passionflower butterflies mimic each other throughout the regions of South America to such an extent that it is not possible to identify a species merely by looking at it. Breeding of *Heliconius* is quite possible and interesting genetic work can be carried out.

Above
Hemaris tityus, the Narrow-bordered Bee Hawk, successfully imitates a bumble bee as it flies by day across the downs, settling to feed on scabious and other flowers. *Hemaris* are found throughout the Palaearctic region and in North America.

Right
The Owl Butterfly (*Caligo*) is one of the banana-feeding Brassolidae, renowned for its curious markings which, on a set specimen, remind one of an owl's head. The butterfly here was hatched from a pupa in England, sent from Brazil and it is a species that might well be bred outside the tropics.

Right
Day-flying moths tend to be more brightly coloured than night-flying ones. This *Dysphania militaris* from south-east Asia is quite commonly found. Its pupa is curiously marked with two prominent black eyes and a projection between that greatly resembles a nose.

Above
One of the most beautiful of European garden butterflies is the Peacock (*Inachis io*) whose red colouring richly contrasts with the foliage it is resting on here. It is not only tropical species that display dazzling colour!

to look at a specimen of, for instance, *Papilio machaon* and know just which part of the world it has come from. Some of the butterflies found in Ireland, though the same species as those found in other places, have quite distinctive colouring and the Common Blue (*Polyommatus icarus*) in Ireland is the brightest and most beautiful form anywhere in Europe. *Papilio polytes* and *P. memnon* have forms which are totally different in colouring, and some with and without tails according to which part of Asia they are found in.

Aberration and natural variation occur in Lepidoptera as in other creatures and collectors make a point of looking out for interesting variations because markings tend to be consistent and aberrations are rather rare. I have some exceptional examples which include swallowtails that are so black that their markings are almost totally obscured. There is also a *Papilio aristolochiae* from India with homeotic variation–that is, a part of the pattern misplaced on to another part of a different wing, rather like a piece of transfer breaking loose and fixing itself on another part of the design. Halved gynandromorphs are some of the most spectacular aberrations–that is, a butterfly or moth which is half male and half female so the left side might be blue and the opposite side quite another colour, such as black.

Habitats and conservation

Butterflies and, to a great extent, moths frequent particular types of habitat and to see as many as possible it is necessary to travel to forest, mountain, grassland or wherever the insects are that you are wanting to see. Their presence is often determined by whether the larval foodplant is there, as much as by other ecological factors. Firstly, we are going to look at some of the most productive types of habitat for butterflies and moths and later consider conservation aspects both of the habitats and of their Lepidoptera.

Below
Papilio ulysses, from Australia and New Guinea, is well known for its bright blue colouring which is only matched by the *Morpho* genus in South America. The butterflies are attracted to the ground for moisture after rain and can be attracted also by blue objects which they take for one of their kind.

Tropical rain forest

The equatorial regions are undoubtedly the most productive and tropical rain forest abounds with Lepidoptera. However, visitors from temperate countries are sometimes disappointed because butterflies are not teeming wherever they look and it is necessary to learn the best spots to find them. The jungle is not a mass of brightly coloured flowers; in fact, it is mainly green with dense foliage, with occasional flowers or groups of them at various levels from the ground right up to the tree tops. Butterflies will usually be found in the vicinity of flowers and certainly around rivers, streams, waterfalls and lakes. Butterflies fly at the edges of forest and along paths or roads; very few like the densely growing areas, which is just as well as these are rather difficult for us to get about in as well! Certain butterflies fly high in the tree canopy though they may come down to feed. I remember well the reward for my patience when I first saw birdwings flying high up like buzzards in the trees in the New Guinea jungle. Eventually some of them came down to feed from the Pawpaw blossom and I counted seventeen flying round one tree!

There are definite flight paths which become apparent when you spend a few hours in the jungle and if you find one it is possible to see scores of different species. At night some butterflies congregate and you might be lucky enough to come across one of these roosting places.

Although South America has perhaps the greatest number of very exotic species there is tremendous variety and interest to be found in any of the equatorial regions of Africa, Asia and Australasia as well.

High altitude

Butterflies often favour high ground and mountainous regions have their own distinctive fauna and flora. Even very high up, above the snow line, butterflies can be found. In North America, certain of the butterflies found otherwise only in the Arctic are to be seen at the highest altitudes. Rare species of *Parnassius* are found in the Himalayas and the Pamirs, south of China and in Afghanistan the mountains provide a number of very interesting species not found outside these parts of Asia Minor. The Alps provide an abundance of butterflies that is perhaps only equalled by the tropics and a range of species and local forms greater than anywhere else in Europe. I shall always remember the thrill of coming across a field of Lucerne on one of my first visits to the French Alps and finding it positively alive with large fritillaries, Large Coppers (*Lycaena dispar*), Large Blues (*Maculinea arion*), several species of Satyridae including the largest in Europe (*Hipparchia circe*), brimstones, clouded yellows and many other kinds; some thousands of butterflies all in one field. The abundance in this region is probably due to the fact that agricultural methods are simpler and many parts are too rugged to be cultivated at all, creating a natural reserve for Lepidoptera and their foodplants.

Grassland

In temperate regions, the downlands and grassy meadows provide some of the best habitats for butterflies which thrive among the variety of plants that favour this type of habitat. This is the stronghold of the blues (Lycaenidae) above all but also a wide range of Lepidoptera from many families, including some lovely day-flying moths. This type of habitat bordering woodland forms the widest range of all types of temperate habitat outside the Alps.

Below
Ornithoptera priamus is one of the beautiful birdwings of New Guinea. The female is larger though not as colourful, being patterned in shades of brown and yellow. Both the caterpillars and the chrysalids of birdwings are exceptionally large and are eaten as delicacies by the natives of New Guinea.

Above
Boloria pales is a butterfly of the high Alpine regions. It is one of the fritillaries that can also be found in the true Arctic. Flying with this butterfly when it was photographed in the French Alps were rare burnet moths, high-altitude satyrids and a mountain form of the Marsh Fritillary.

Right
The brightest of the grassland blues is *Lysandra bellargus*, the Adonis Blue. It is a species that is fast disappearing and reserves are being used to encourage the butterfly to increase in numbers again.

Woodland

As with the tropical rain forest, butterflies are not found deep in woodland but rather along the perimeter and along rides or in clearings. Coniferous woods are less productive than deciduous, but a lot depends on the variety of vegetation that grows in the clearings and the amount of sunshine that is able to penetrate. Brimstones (*Gonepteryx*) are the first woodland species to appear in spring, after hibernation, followed by Vanessidi such as Peacocks and Tortoiseshells.

Left
A much prized species is the woodland Brown Hairstreak (*Thecla betulae*). The larvae feed on Blackthorn and are protected by their resemblance to the foodplant.

Below left
One of Europe's grand butterflies of woodland is this White Admiral (*Limenitis camilla*). The caterpillar hibernates in an individual hibernaculum which it spins on honeysuckle, and the butterfly particularly favours bramble blossom.

The range of butterflies found in woodlands has similarities right across the Palaearctic region (stretching from Europe across Russia to Japan) and even across North America. Fritillaries are characteristic of woodland, both *Argynnis* and the allied *Melitaea* and *Melicta* sometimes flying in quite large colonies and feeding avidly on the blossoms of bramble or thistle flowers. Powerful Nymphalidae such as the purple emperors (*Apatura*), and the admirals (*Limenitis*) soar high up in the trees, coming down to feed from moisture, flowers or the sap of trees. In the undergrowth a number of Satyridae will be found where they live on the grasses, sometimes with blues and coppers. Woodlands usually have a number of hairstreaks (*Thecla* and allied genera) breeding on the trees and the shrubs below them.

Hot and cold deserts

Nature has evolved ways in which both animals and plants can overcome extremes of climate and butterflies are to be found even in very adverse conditions. In exceptional drought some insects remain dormant and then emerge when the rains come. Hesperiidae, the skippers, are often to be found in desert localities, together with several nymphalids, danaids and even sometimes papilios. The colours tend to be dull or sandy so they blend with their surroundings. The Arctic produces quite an array of satyrids, fritillaries, clouded yellows (*Colias*) and even *Parnassius*, but all these are absent from the Antarctic.

Above
The Glanville Fritillary (*Melitaea cinxia*) is found in Britain only in the Isle of Wight. Although widespread in the rest of Europe, its range is so limited in Britain that it could disappear easily and measures are being taken to conserve it.

Conservation

In recent years conservation has become very important and a new approach to wildlife is necessary. It is a new science and little is yet known about the needs of Lepidoptera to ensure their future survival. Books on Lepidoptera have usually been concerned with their study from the point of view of a collector and even recent works omit details of the butterflies and moths as living creatures. By studying the life histories and getting to know the species thoroughly we have the opportunity of maintaining the habitats as they need them; otherwise the decline will surely continue.

There are two sides to the conservation issue. There are people who are not interested enough to bother about the future survival of wildlife or

Below
Parasites take their toll of butterflies. This White Admiral caterpillar is almost covered with cocoons of an *Apanteles* species of parasitic fly which feeds on the inside of the caterpillar before emerging to pupate outside its skin.

those whose immediate interests in development take a higher priority. On the other side there are the reactionary conservationists who put conservation above all other considerations. It is easy to be swept away by a cause. Either one of these two extremes is unlikely to be the best approach for the future of both the human race and other forms of wildlife. There must be maximum cooperation between the conservation bodies and those who are concerned with the development of land and, if both steer a moderate course, each will succeed in their aims. Conservationists must give some consideration to the needs of man for agriculture, forestry, housing and communication and those who are concerned with development should take into account the needs of wildlife and help to create areas where creatures and plants can continue undisturbed. It may well be possible to move the course of an intended road development but it is just as futile for the conservationist to expect the road building to be stopped altogether as for the developer to think he can drive roads through the countryside with regard only to his own convenience. Fortunately such cooperation is resulting in both development and good conservation but there is a long way to go yet and there is no room for hotheads!

The greatest single factor in the decline of Lepidoptera is the destruction of natural habitat. The use of insecticides and herbicides is sometimes suggested as the primary cause but this is not as damaging as the total destruction of a wild piece of land for development. Even the jungle is being eroded away by conversion to plantations of banana, rubber and so on, and for road building. Climatic effects must also be taken into account. A recent fall in the British population of the Large Blue butterfly, which is now almost extinct, was caused firstly by a season which was so wet and overcast during the flight season that the butterflies hardly laid any eggs and then by such a sunny, dry summer that the thyme plants dried up and the butterflies lived only a few days and again few eggs were laid.

I am a great advocate of studying butterflies and moths alive and watching them both in their own environment and one's own conditions at home, but I am not against collecting Lepidoptera and do not believe it is attributable, to any perceptible extent, for the disappearance of any species. Above all, while I hope that in the future it will be possible for us to show children how to begin studying butterflies without collecting them, it is a way in which many children come to learn about Lepidoptera and they enjoy chasing about in the countryside. Many of us started our interest in entomology in this way and it need not be condemned; rather it is better to guide a young person into more productive studies than to prohibit him from collecting. Furthermore, for true scientific study collecting is essential. Much of the original classification work has been done by private collectors and amateur entomologists and continues to be. Identification in many cases can only be made by dissection and we learn a great deal about the relationships of Lepidoptera

Left
One of the rarer hairstreaks of Europe, the Black Hairstreak (*Strymonidia pruni*) is in danger of disappearing from Britain. It is being closely studied and its habitat carefully maintained. The larvae feed on Blackthorn but the butterflies favour only ancient uncut growth.

Above
The Map Butterfly (*Araschnia levana*) is found in Japan and Europe but, although it may once have been native to Britain, it is not found there now. An attempt to introduce it was spoilt by an entomologist who destroyed a colony that had been established in the Forest of Dean.

by studying collections. At school we used to run a moth trap every night. From this we formed a most useful reference collection, learnt a great deal about the moths and learnt to identify many hundreds of species. Having kept the specimens we needed for the collection and some females for egg-laying, we would release most of the catch into undergrowth away from birds.

The loss of the Large Copper (*Lycaena dispar*) in Britain has been attributed in a very recent publication to over-collecting but the author ignores the fact that its disappearance coincided with the drainage of the fens on which the butterfly relied for its correct habitat! Disappearance of a species can practically always be traced to destruction of habitat. There is a concern about the amount of collecting that is carried out on the island of Taiwan where millions of butterflies are caught every year. The concern is justified because we should always be aware that what we do today could be harmful tomorrow. However, I have seen how very prolific the butterflies are in Taiwan and know that collecting on this scale has continued over more than twenty years, yet the butterflies are back again each year. There appears to be some natural balancing of nature. Again, destruction of habitat turned out to be the cause when it appeared that three Taiwan butterflies were becoming scarcer; their breeding grounds had been bulldozed to make way for hotels. Unnecessary exploitation should certainly be avoided but some collecting can be permitted without being a conservation hazard. Many collectors are now turning to photography to record their entomological expeditions. This and the study of butterflies alive is certainly to be encouraged and is, of course, essential when the species involved is in any danger of extinction.

There is a little more to establishing a butterfly colony than simply releasing a number of butterflies into the wild. Often the conditions are not quite as they need to be and the attempt fails. Very little study of the necessary ecological conditions has been made until recently and this is where amateur entomologists can play a very valuable part. We need to watch Lepidoptera in the wild and record their requirements. Above all, a constant watch needs to be kept on the population levels with notes made of the fluctuation conditions. This will guide us as to the conditions in which Lepidoptera thrive and enable us to introduce species into the wild, or strengthen existing colonies, with greater success. The British Butterfly

Conservation Society runs a habitat survey scheme which records just these details. The scheme is unique so there is an open field for similar research in other parts of the world. Observations are sent in from members, most of whom are amateurs, and the results of their work will be most valuable to science in the future. The Society was founded in 1968, and run for several years just by my wife Rosemary and myself. The habitat surveys were introduced shortly afterwards and now the whole scheme has grown and many hundreds of members contribute their work. Such societies are few but can be started in other parts of the world and they will do a great deal to engender interest in conservation.

A division of the Nature Conservancy Council, in Britain, runs a population survey of Lepidoptera which is linked to a similar scheme in other parts of Europe. Observers, who are mostly amateurs again, send in reports of the species they record in designated 10-kilometre squares and from this an accurate record is plotted on a computor showing the distribution of the entire range of Lepidoptera. The Biological Records Centre welcomes enquiries from any enthusiast who would like to take part in the recording scheme.

Those who would like to help in the conservation of Lepidoptera would be well advised to join a local or national society. It may be that there will be recording schemes that you can take part in and experts from whom you can learn.

Helping in conservation education, particularly in schools, is a valuable contribution and if you have any appropriate knowledge so that you can give talks to societies you can count on considerable interest.

Having looked at observation and recording there are less scientific aspects, of a practical nature, to be investigated. Many people are interested in attracting and maintaining plenty of butterflies in their garden and to some extent this is possible. Everything depends on having a great abundance of flowers, preferably throughout the period of the year that butterflies can be expected in. Butterflies depend on nectar and it is flowers that attract them, so if your neighbour has more flowers than you, the butterflies will be attracted to his garden. If you have the space you may wish to go a little further and encourage butterflies to breed and this entails allowing parts of the garden to grow a little wild (and saving yourself some trouble in maintaining it!). Certain flowers are known for their ability to

Left
Marsh Fritillaries (*Euphydryas aurinia*) live mainly on Devilsbit Scabious, though in captivity they thrive on honeysuckle. They can be established rather successfully in suitable habitat and seldom stray if conditions are to their liking.

Above
Some butterflies are found only in very restricted localities. The Provence Short-tailed Blue (*Everes alcetas*), from France is one of these. It is seldom photographed, but fortunately it presented itself unexpectedly on a recent visit to the Alps.

Above
The Ice Plant is one of the most successful of all plants for attracting garden butterflies. It is important not to ask for the more vigorous and darker variety 'Autumn Joy' as this does not have the same effect.

attract butterflies and it is best to try to grow these. Here is a list of some that can be grown for butterflies in temperate countries:

Alyssum
Aster (*Callistephus chinerises*)
Aubrietia
Buddleia
Bugle (*Ajuga reptans*)
Catmint (*Nepeta cataria*)
Cornflower (*Centaurea cyanus*)
Echium
Golden-rod
(*Solidago virgaurea*)
Heliotrope
(*Heliotropium europaeum*)
Honesty (*Lunaria*)
Ice Plant (*Sedum spectabile*)
Lavender (*Lavandula*)
Michaelmas Daisy
(*Aster novi-belgii*)
Mignonette (*Reseda odorata*)
Phlox
Polyanthus
Sweet Rocket (*Hesperis*)
Sweet William
(*Dianthus barbatus*)
Thrift (*Armeria maritima*)
Valerian (*Kentranthus ruber*)
Verbena
Wallflower
(*Cheiranthus cheiri*)

As well as attracting wild butterflies you can breed your own for release. It is possible to buy chrysalids and hatch them out and you can breed your own from wild collected stock. Details of how to breed Lepidoptera are outlined in the final chapter of this book.

It is difficult to create an artificial environment for Lepidoptera and it is preferable to maintain or develop an existing habitat. If you have in mind a particular piece of land, have a look at it first and find out what already exists there. I once surveyed an entire country estate for an owner who wanted to encourage butterflies but it was either so exposed or over-cultivated that it was clear that most species would not establish themselves there. It may be, however, that you merely need to encourage the spread of the foodplant or provide more nectar-bearing flowers. Watch out for excessive undergrowth as this sometimes takes over a piece of land which was once ideal for butterflies and chokes them out of existence.

Principal families of butterflies and moths

Below
This hesperi d, *Odontoptilum*, from Hong Kong, displays disruptive coloration which could also be assumed to be imitation of the familiar bird-dropping pattern.

The binominal system of naming was introduced by Linnaeus in the eighteenth century and Latin or Greek words usually provide the names. These are international and used by zoologists regardless of their own language. The first name is that of the genus (expressed with an initial capital letter) and the second is the name of the species (written in lower case letters). Thus we have *Papilio machaon* and if we wish to refer to a particular subspecies its name is added so we get *Papilio machaon britannicus*.

Lepidoptera are classified into families, genera and species according to sets of physical characteristics. A particular wing-shape, together with a standard pattern of veining may be the characters which place a species within a particular family. In the survey of the most prominent families that now follows, some of the important features that place the species in to the family in question will be outlined.

Hesperiidae Skippers are rather moth-like in appearance. They have a swift, darting, or even buzzing, flight and many hold their wings in a moth-like position when at rest. They are thought to be the most primitive family of butterflies. An exceptional species in Australia, *Euschemon rafflesia*, has the bristle and catch mechanism that normally defines a moth. Some taxonomists rate the Hesperiidae as a superfamily of its own, which gives us butterflies, moths and skippers, all of an equal status! The size is usually small, average European species being about 2 centimetres (0·8 inches) across; tropical species can be twice this size and there are certain exceptional species which measure about 9 centimetres (3·5 inches) across. Skippers occur in all geographical regions. There are about thirty species in Europe but throughout the world the number of species runs into many hundreds. A good many feed on grasses in the larval stage. The caterpillar draws the blade round itself and lives inside this tunnel. So as not to attract attention to itself by accumulating piles of frass (excreta) at the base of the plant the caterpillar catapults each piece away! Some skipper larvae have large spots on the head which are like greatly enlarged

eyes, sometimes in bright colours, which give them the appearance of being fierce. The majority of skipper butterflies are not brightly coloured, usually a shade of dull brown or orange-brown, but there are some colourful exceptions in Asia and especially South America, where several of the species have long, sweeping tails.

Papilionidae This is a very varied and well-known family of butterflies, represented by over 500 species throughout the world and noted for their size and exceptional beauty. In Britain, the Swallowtail (*Papilio machaon*) is the sole representative and over the whole of Europe there are only about a dozen species, of which fewer than half are swallowtails as these include *Parnassius* and allied genera. In fact many of the true swallowtails are tailless and, such is the variation in appearance, there seems to be no comparison at all between Apollos (*Parnassius*) of the Alps and the giant East Indian birdwings (*Ornithoptera*) although they are in the same family.

The majority of Papilionidae are in the tropics and the jungle vines (*Aristolochia*) are selected as the foodplant for whole groups of species, especially in South America and in the Indo-Australian regions. The larvae show a great family resemblance, being black or dark, especially in the early instars, with many fleshy tubercles and a characteristic white 'saddle' on the back. This type of caterpillar is found in the *Ornithoptera* as well. Most of the other Indo-Australian *Aristolochia* feeders are black with markings often in red and these colours warn predators of their poisonous nature. Likewise the 'black papilios' (*Parides* and allied genera), of South America, have these colours but here there are additional greens and iridescent blues which make them most attractive to look at. There are also groups of swallowtails, both in the Old and the New World, which have *Citrus* (orange, lemon etc.) as their foodplant and the third family of plants which are a particular favourite of whole sections of the

Left
This is the Lulworth Skipper (*Thymelicus acteon*). A more typical resting position for a skipper would be the hindwings lying flat while the forewings are held almost vertically above them.

Above
Papilio anactus from Australia is found both in the tropical and subtropical regions. It is a common species found especially wherever *Citrus* grows.

Papilionidae is the Anonaceae (custard apple family). The swordtails or *Graphium* genus live largely on these plants.

Papilionidae are among the most long-lived of butterflies. In the wild they may live for two to three months where it is not excessively dry and there are plenty of flowers. They will drink from moist river banks but seldom feed from sweet secretions and they do best with plenty of nectar flowers, sprayed occasionally with plain water. With a little experience of breeding butterflies certain of the swallowtails are not too difficult. The problem with many butterflies is in providing the correct conditions for them to pair and lay eggs. The Papilionidae are easier to hand pair than most. This is a process by which the male and female are held together in such a way that they join the tips of their abdomens, in conditions that do not persuade them to do this naturally. They part after an hour or so and the female then lays normally. *Papilio machaon*, which feeds on fennel and related umbellifers is one of the easiest to breed, and so are related species such as *P. polyxenes* and *P. zelicaon*. *P. demodocus* from Africa, and its Asian cousin *P. demoleus* breed well in captivity on *Citrus* and it is possible to breed strains that feed on fennel. Little breeding work has been done on the *Aristolochia* feeders outside the tropics but, in heated conditions, it might well be possible to open up a whole new field here.

The *Parnassius* or Apollo butterflies are associated with

high altitude. They are common on certain European mountains where you cannot fail to notice them for their curious white, papery appearance as they float about in the air. They are marked with prominent black spots and dark veins, usually with red discs or circles. In parts of the Himalayas and Pamirs there are rare species, sought by collectors. *P. autocrator* is one of the rarest, with bright patches of yellow on the hindwings, not found on any other species.

Next in the chain of evolution, with characters of both *Parnassius* and *Papilio*, are butterflies from Turkey, parts of Europe and China which include the tailed *Sericinus*, *Doritis* which is very Apollo-like, *Helios* and other genera which were known as *Thais* but have now been split into several new genera. To see these ranged in a collection is particularly interesting as they show this evolutionary relationship very clearly. One of the most beautiful of all Papilionidae, *Bhutanitis lidderdalei*, comes in this chain. It has extraordinarily patterned, long, narrow wings in black and white, with multi-tails and a vivid patch of red with yellow and blue on the hindwing. This butterfly comes from the mountains of northern India and has a rarer cousin known as *B. thaidina*.

The swordtails and other, non-tailed, members of the genus *Graphium* are important among the Papilionidae. The European Scarce Swallowtail (*Iphiclides podalirius*) is closely allied and the largest species, with very long, pointed, sword-like tails, is *Graphium androcles* from Celebes which is a very impressive butterfly, strongly barred in black and white. *Graphium sarpedon*, the Green Triangle or Blue Bottle (one can see why international scientific names have been adopted!) is one of the commonest and it is found throughout the Indo-Australian region. There are several quite distinct geographical forms. *Graphium doson*, *G. evemon* and the lovely lime-green dappled *G. agamemnon* are often found in the same parts. In Africa there are more graphiums than other papilios. Most are not tailed and few are bright but they include rarities which are of great interest to students of African Lepidoptera. Several fine swordtails are found in North America, one of the prized ones being *Graphium marcellus*. Equatorial South

Above left
The Scarce Swallowtail of Europe (*Iphiclides podalirius*) was so named because of its rather rare appearances in England. In other parts of Europe it is not uncommon and is a wonderful sight as it drifts and almost sails in the wind. The Germans call it the Sail Butterfly.

Above
Leptocircus curius is closely allied to the swordtails but the wing shape of this genus is unique and very exaggerated. The flight of this butterfly, which comes from India and south-east Asia, is peculiar for its very fast wing beat, making it resemble a skipper (Hesperiidae) as it buzzes from flower to flower.

Above left
This Malaysian swordtail is *Graphium antiphates*. It is found in rain forest clearings and on river banks through Borneo and the Philippines across to Hong Kong. The butterfly congregates in damp places, with other species, sometimes in hundreds, to drink the moisture from muddy banks.

Above
Entomologists visiting Corsica and Sardinia always hope to catch a glimpse of this relative of the commoner *Papilio machaon*, for *Papilio hospiton* eludes most people and is usually only seen in rather inaccessible mountainous country.

Left
This handsome Moss Green Swallowtail (*Papilio arcturus*) is found in India in the hills of Assam and Nepal. Its life history is so far unknown.

America has a range of some ten or more black and white barred species which are, therefore, difficult to identify. *Graphium thyastes* is noted for its beautiful rich, orange-yellow colouring which makes it stand out from the myriads of others when feeding in clusters on river-banks, a common habit of graphiums both in South America and in Asia.

Two very striking Papilionidae in Africa might well be taken for one of the birdwings. They are in different genera but they are both noted for their extraordinary size and build. Little is known of their life-history and the females of both are hardly ever seen. *Iterus zalmoxis* has a bright yellow, very fat body (just like the birdwings). Its wings are a beautiful gunmetal blue, rayed with black. *Druryeia antimachus* has wide, narrow wings rather like a heliconid but with a wingspan equal to that of the Ornithoptera. It is coloured with the characteristic African warning colours of orange-brown and nearly black, and it has sufficient poison in it to kill even quite a large bird or mammal if eaten by one.

The Indo-Australian region has a group of very beautiful moss green and blue swallow-tails. The finest blue species, *Papilio ulysses*, has the iridescent blue brilliance of the South American *Morpho* butterflies, banded with black, and with beautiful rounded tails. *Papilio montrouzieri* from the Philippines is like a miniature of *P. ulysses*, and *P. pericles* from Timor has a little purple added. The largest is the magnificent *P. blumei* from Celebes, which has brilliant flashing bands of green which merge into iridescent turquoise and become a shining sky blue on the boldly rounded tails. Three other iridescent green species are particularly prominent: Sri Lanka has *Papilio crino*; *P. palinurus*, with very bright bands, comes from Malaysia and the Philippines; and *P. peranthus* with an entire green centre to the wings, is found in Java. A fourth, *P. buddha*, from India is one of the species collectors are always hoping to obtain but it is seldom

Above
Papilio memnon is a large swallow-tail occurring throughout tropical Asia. It is variable according to its geographical location. In some parts it is coloured with red, in other regions orange and elsewhere yellow. In addition to this some forms are without tails. This is the female; the male is usually tailless and coloured steel blue on black.

Right
Parnassius are mountain-dwelling Papilionidae that are not actually swallowtails. *P. apollo* from Europe has the typical pattern of the group, which occurs in all cold regions, usually at high altitudes. The larvae usually feed on succulent plants such as *Sedum*.

Left
Between *Parnassius* and *Papilio* come a group of butterflies that are intermediate in their evolution, of which this *Alancastria cerysei* from Bulgaria and Turkey is an example. Others, which used to be in the genus *Thais*, but are now divided into new genera, occur in Europe and parts of Asia Minor.

Above
The *Delias* are Pieridae, renowned for their beautiful underside markings. *D. eucharis* is from India and other species come from various parts of tropical Asia. Their stronghold is in the mountains of New Guinea and on other islands in the East Indies. The larvae of most species feed on species of mistletoe.

seen in collections. There are some twenty other beauties in this group, including *P. paris* which is quite common, and *P. karna*, which rather resembles *P. paris* with its bright round pools of green on the hindwings, but is quite a rarity.

The birdwings of New Guinea and Australasia are rated by many as the most glorious and impressive of the world's Lepidoptera. This must be due not only to their size but to the shining colours and patterns on their improbably shaped wings. The females, which are not brilliantly coloured like the males, are rather similarly marked in shades of brown and their size is breathtaking, especially so when one looms down towards you out of the trees, larger than most of the birds! *Ornithoptera paradisea* has the most exotically shaped bulbous forewings and the contrastingly shaped hindwings taper to such a fine, long point that the tip usually curls into a mere whisp. The colour is a heavenly iridescent lime green, bright, shining golden yellow and black. Only a live specimen conveys the excitement of the true colours–I have never seen a picture that can do so. The rarer *O. meridionalis* is a little smaller but with a rather similar pattern and wing shape. *O. goliath* is a sheet of the wonderful colours just described and very large. Several other extraordinary species from New Guinea are in this group and one of these, *O. alexandrae*, must be mentioned, as the female is the largest butterfly in the world and the male has wings that are strangely shaped, like four great lobes rather than wings, and coloured with an exciting combination of shot green mingled with blue. Many, less rare, species of black and yellow birdwings are found throughout the region including the famous Rajah Brook's Birdwing, *Trogonoptera brookiana*, of Malaysia which has a slender wing shape and bold triangles of iridescent green, spaced across its velvet black wings. It is sometimes seen in flocks of fifty or more, drinking from the muddy river banks.

Pieridae This colourful family of butterflies is characterized by yellow and white with markings of black but many have, in addition, brilliant patches of pinks, oranges, reds and greens. In Europe the pierids include the Large White, Orange Tip, Brimstone, clouded yellows, and several other whites. In North

Above
Among the African Pieridae, *Pontia helice* is one of the most attractive. This common species, like the *Delias*, has its most brilliant colours on the underside. The upperside is black and white, not unlike the European Bath White (*Pontia daplidice*) to which it is related. This picture shows the male and female pairing.

Left
The beautiful underside of *Catopsilia scylla gorgophone*. This pierid from Australia has white forewings and hindwings of bright yellow, as if two halves of different butterflies had been pieced together.

Above right
One of the largest pierids in the world is the Giant Orange Tip (*Hebomoia glaucippe*). It is found from India eastwards to the Philippines where related species, some with beautiful citron yellow colouring, have evolved.

America there are counterparts to these, and likewise right across the Palaearctic region to Japan. In the tropics some of the orange tips and brimstones are giant versions of those we meet in temperate regions. Pierids stretch right up into the Arctic where there are some very interesting endemic clouded yellows with curious moss-green colouring. There are also pierids that withstand the rigours of parched lands in Asia Minor, such as the *Euchloë* (cousins of the orange tips) and *Colias* which also includes species occurring high up in the mountains of Iran and Afghanistan that are seen nowhere else. Certain Pieridae breed easily in captivity but they need flowers rather than moist pads for feeding. They succeed best if given growing foodplant to lay on.

Central and South America have three large and beautiful brimstone (*Gonepteryx*) species. *G. menippe* is richly coloured in yellow, with broad orange tips. *G. clorinde* is the palest whitish green with two splashes of yellow on the leading edge of the forewing, and *G. maerula* is vivid yellow all over, like the European Brimstone (*G. rhamni*) but more than twice its size. True clouded yellows (*Colias*) are not found in the tropics. There are several species in North America and a closely allied genus which has the famous Dog's Face Butterfly (*Megastoma caesonia*), so named because its markings are like a silhouette of a poodle's head. There are *Colias* in the Andes mountains and otherwise the best species are found in the rather inaccessible places mentioned a little earlier. True orange tips (*Anthocharis*) are found only in the Palaearctic region and in North America where there are species of exceptional beauty. In the tropics the giant *Hebomoia* of Asia are many times the size of *Anthocharis*, and although their relationship is not very close, their resemblance is. Africa has a group of tipped butterflies, the *Colotis*, many of which are like orange tips to look at but there are also species with tips of purple and red.

Some of the most colourful of the world's butterflies are the *Delias*, found all over Asia and especially Australasia. The rarest are mostly found on rather inaccessible mountains in New Guinea. It is the underside of *Delias* which is so splendidly coloured in extraordinary patterns of red, yellow, chocolate and orange. The uppersides are usually black and white–rather reminiscent of a white.

The *Catopsilia* butterflies must be mentioned. These are yellows, often of great intensity of colour, and are found in all hot countries, from the Mediterranean southwards. They often migrate in quantity, sometimes, it seems, to move from an area that is too dry and certainly in order to disperse the species. One of the most prized species, *C. avellaneda* comes from Cuba. It is yellow, with red suffused markings, almost as if it were stained with blood.

Above
Fritillaries are noted for the silver spotting on the underside. This High Brown Fritillary (*Argynnis* (*Fabriciana*) *cydippe*) is a woodland butterfly, sometimes seen in some numbers, especially when thistle flowers are present.

Right
Kaniska canace is a vanessid from China and Japan, related to the Large Tortoiseshell (*Nymphalis polychloros*) and with the same woodland inclinations. It extends southwards to subtropical regions and this specimen comes from Hong Kong.

Nymphalidae Sometimes these are referred to as the brush-footed butterflies because one of the principal criteria for their classification is that they have only four walking legs, the other two being degenerate and formed like tiny brushes. The Nymphalidae embraces several distinct groups of butterflies including the fritillaries, emperors, vanessids, *Agrias*, *Catagramma* and many others. Modern classification includes butterflies which have always been regarded as distinct families of their own, such as Danaidae and Heliconiidae but here they will be treated as distinct to avoid adding to the already overcrowded miscellany of the family Nymphalidae! Most species are very swift, powerful fliers; a quick flap and a glide, or a magnificent burst of speed up into the trees is the typical flight of the nymphalid. Feeding takes place on practically anything available and, in captivity, nymphalids will take their nourishment well from honey pads.

The Vanessidi are some of the most familiar butterflies in gardens of temperate countries. These include the Peacock, the Small Tortoiseshell (*Aglais urticae*), Red Admiral, and the Painted Lady which seems to migrate to practically every country of the world! Some of these butterflies rely on Stinging Nettle as the sole foodplant of the caterpillar and can be encouraged by allowing Nettle patches to grow and not over-tidying garden perimeters. Vanessidi include woodland butterflies, the large tortoiseshells, commas (several species across Europe, Russia, Japan and America) and some species in Australia and New Zealand. Closely allied to the Vanessidi are the *Precis* or pansy butterflies. Several species have attractive eye spots and there are quite a few that can be reared in captivity. Pansies are found especially in Africa, Asia and North America but not at all in Europe.

Fritillaries are named after the spotted Fritillary flower and they include several genera, all of which have the same general pattern of dark-coloured spotting on orange-brown. On the underside there are usually spots or splashes of silver. They are mostly associated with woodland but some species live in field and downland. The largest is *Argynnis diana* from North America, the female of which is unusually coloured in steel blue and black. North America has the most species of fritillary with certainly two more that are as spectacular as *A. diana*. These are *A. nokomis* and *A. idalia*. A very fine and large fritillary, perhaps the most beautiful of all, *A. childreni*, comes from northern India. It has a beautiful rich pink flush over the underside forewings, which contrasts spectacularly with the aluminium silver and moss green colouring of the rest of the wings. There are endemic Arctic fritillaries and some Alpine species that are of interest to the specialist. Many *Argynnis* larvae feed on violet and the field-dwelling chequer spots (*Eyphydryas*, *Melitaea*, *Melicta* etc.) live, often communally, on plantain, scabious, mallow and other field plants.

Other woodland nymphalids of temperate countries include the Purple Emperor (*Apatura iris*) and the related *A. ilia*. The rich purple colouring of the male cannot fail to impress. Flying with *Apatura* may well be admirals (*Limenitis*) and the related *Neptis*. This latter genus has a delightful gliding flight that immediately identifies it. The genus is represented in tropical Asia, Africa and North America as well as in Europe.

A very impressive genus, which has its headquarters in Africa, is the *Charaxes*. Only one species occurs in Europe, a very beautiful, tailed butterfly with the characteristically contrasting patterned underside of *Charaxes*. A few *Charaxes* live in tropical Asia and the rest are in Africa. They can be attracted by using bait of rotting fruit and quite a variety of species may come at any one time in the right season. A great many are coloured in shades of brown but *C. bohemani*, one of the larger species, is a lavender blue, the rather small *C. eupale* is a beautiful leaf green and others, such as *C. zoolina* from Madagascar, are white.

The much more brilliantly coloured *Prepona* and *Agrias* of South America have much the same robust build as *Charaxes* and the same habits of fast flight and attraction to carrion and fruit. *Prepona* are most easily identified from their undersides which have an infinite variety of patterning like a dead leaf. The uppersides are usually banded with iridescent sky blue on black and some are shot with violet. However, the colouring of *P. praeneste* and its close relative *P. buckleyana* is more like that of an *Agrias* species, with bands of scarlet and pools of iridescent purple. When you see a collection of *Agrias* you realize that the birdwings and morphos are not the only beauties in the world! The brilliance of some of the iridescent purples, magentas, greens, blues and reds, contrasting with bold outlines of black, are quite impossible to imagine. Rather little is known of their life history and most are obtained by trapping with bait.

Heliconiidae These live almost exclusively in tropical America, only one species, *Heliconius charitonius* reaching as far north as Mexico and the southern United States. They are medium sized, with long, slender wings. They are extremely numerous and there are dozens of species. Some are very gaily coloured with scarlet, blue and yellow on black. Heliconiids are renowned for their mimicry of each other and of other families which resemble them in appearance so effectively that it is not always easy to distinguish the species without dissecting the genitalia. This is a most interesting family to study and a collection of all its species and variations would be very large indeed.

The larvae feed on passion flowers. A final instar larva is a magnificent sight with six spines branching from each segment. *Heliconius* can be bred in captivity in the right conditions with growing foodplant and plenty of flowers. The pupae are curiously shaped, sometimes with long horns or decorated with rows of barbed spikes.

Morphidae These magnificent butterflies compare with the East Indian birdwings in their spectacular size and brilliant colouring. They are particularly known for the metallic shades of blue that colour the males, colours which must be seen to be fully appreciated. Morphos live only in Central and South America and the majority occur in the tropical regions of Brazil, Peru, Bolivia and Colombia. While *Morpho aega* is a little over 10 centimetres (4 inches) across, the largest, *M. hecuba*, can have a wingspan of 25 centimetres (10 inches). *M. hecuba* is one of the

Left
The Red Underwing Moth (*Catocala nupta*) is a member of the very large family Noctuidae. The brightly coloured hindwings are concealed while at rest but on taking flight they are exposed and show a bright flash of colour. The majority of Noctuidae are not as large (they range from 1–3 centimetres, 0·3–1·2 inches across). They are mostly coloured in shades of brown and these form the majority of the kinds that buzz around lighted windows.

Below
The caterpillar of *Morpho achilles violaceus* from Brazil. Newly hatched, it has a curious ruff of rounded black hairs, like eyelashes, surrounding the head. Later, the head is held hanging downwards, accentuating this feature. Eventually the larvae become beautifully patterned and tufted in a way that is quite unusual among butterfly caterpillars.

species of *Morpho* that is not blue, and at least three other species, *M. hercules*, *M. theseus* and *M. amphitrion* are coloured mainly with brown with suffusions of green, silver or orange-brown. They are, nevertheless, by no means dull. The giant *M. hecuba* and its blue form *cisseis* are a splendid sight indeed. There are three species that are translucent papery white or very pale eau de Nil, quite unlike any other butterfly–*M. laertes*, *M. catenarius* and *M. polyphemus*. There is a further group which we might loosely term the *achilles* group, *M. achilles* being the best known of those with this type of colouring which is basically black with a very broad band of iridescent blue or violet running down both the fore-wings and hindwings. In *M. peleides* this band covers most of the wing. The underside of the *achilles* group is beautifully marked with rings and streaks of green and red on black. The species that have the characteristic iridescent blue include the smaller species, *M. aega*, *M. adonis* and *M. aurora*. *M. menelaus* is a little larger than these, and it has a much larger form, *nestira* which is very like the biggest of the blue species, *M. didius*. Both *M. cypris* and *M. rhetenor pseudocypris* are prettily banded horizontally with white. There are some very attractive species which are a pale, opalescent blue. These include *M. sulkowski* from Colombia, *M. eugenia* and a magnificent large species, *M. godarti* from Bolivia. Morpho larvae, unlike most butterfly larvae, are hairy and they often live gregariously. They are seldom bred and few of their life histories are properly known.

Satyridae The browns are found in all geographical regions of the world and many are the commonest butterflies found in their particular area. They are usually coloured in shades of brown or orange-brown, though a few species are grey or black and white. One of the characters that determines a satyrid is the thickening of the wing veins where they join the thorax. Another common feature is a row of pupilled eye spots or an eye on the forewing. The larvae are mostly grass-feeders and typically have pointed head and tail ends. The pupa is generally suspended but some species pupate on or just under the ground.

Lycaenidae This family comprises thousands of species and forms but it can be loosely divided into the coppers, blues

and hairstreaks. The coppers are the least numerous; their fiery burnished copper colour is unequalled in any of the other Lepidoptera. In Britain there is now only one species, *Lycaena phlaeas*, the Large Copper having become extinct. In the rest of Europe, however, there are a score or more of species and forms.

Coppers occur in all geographical regions but they are less numerous in the tropics. The most numerous in all regions are the blues. In Europe the small blue butterflies which swarm on hillsides and chalk downs in summer are well known (though not as common as they used to be). They are subject to great variation and have been given hundreds of variety names by collectors, who sometimes specialize in collecting just one or two species. The males are generally a shade of blue but the females are brown. In Britain there are eight species, but about ten times this number are found in the rest of Europe.

Throughout the world the blues are nearly always quite small, less than 2 centimetres across (0·8 inches), indeed, the smallest butterflies in the world are blues but in Africa one or two species are about 6 centimetres (2·4 inches) across. Some of the blues are without any blue colouring and in the tropics many are white with black markings. Many are tailed and could be confused with hairstreaks. The undersides are nearly always marked with rows of tiny spots and rings on a pale background and it is these markings that are subject to such variation, referred to earlier.

Above left
Satyridae are mainly grass-feeding species, found on every continent. This Marbled White (*Melanargia galathea*) is very common in parts of Europe. Its colours however are not typical, as most are in shades of brown, usually with little eye spots.

Left
Melanitis leda is one of the larger satyrids, found throughout tropical Asia and Australia. It flies at dusk as well as by day, sometimes breeding in colonies of hundreds covering just a small patch of lush-growing grass.

Above right
The typical shape of a lycaenid larva is illustrated by this caterpillar of the Brown Hairstreak (*Thecla betulae*). The caterpillar is hard to find as it is so similar in size, shape and colouring to the leaves of Blackthorn on which it lives.

Right
Among the European Lycaenidae, this *Heodes virgaureae* is one of the most beautiful with its fiery copper colour. In mountain localities the butterfly is often remarkably common. Coppers are found also in America, Japan and Africa.

The blues are often particularly abundant in tropical countries. They fly in grassy places at the edges of woodland and in gardens. There are very many species in Asia and Indo-Australia and in Africa they are moderately well represented. In South America there are rather few blues but this lack is made up for by a magnificent range of hairstreaks.

The larvae of Lycaenidae are shaped rather like a woodlouse. Generally they are a shade of green to match their foodplant. The larvae of blues feed on plants in the family Leguminosae (peas, clovers and vetches) as a rule. They attract ants which 'milk' them for a honeydew secretion produced from glands on the back. The Large Blue caterpillar actually lives in the nest of a red ant, feeding on the ants' larvae.

The third group, the hairstreaks, is very widespread and includes some of the world's prettiest butterflies. In Europe they are small but some of the tropical species can be up to 6 centimetres (2·4 inches) across and with tails of nearly this length! The colours vary tremendously. In Japan there are species with an overall, iridescent light bottle-green colouring. Many are shining purples and blues and most have undersides that are quite as attractive as the upperside. The larvae are similar in shape to those of the blues and coppers and the majority feed on the leaves of trees, pressed tight against the leaf, camouflaged to match the veining.

Nemeobiidae The metal marks have been known scientifically in recent years by the names Riodinidae and Erycinidae as

Below
The emerald moths are in the vast family of Geometridae, whose caterpillars inch themselves along with a looping motion. The pale green of this *Geometra papillionaria* and other emeralds fades very quickly on exposure to light. The caterpillars match the buds of birch that they feed among, and even change colour to match them in the spring!

Above
The Green Hairstreak (*Callophrys rubi*) breeds among gorse and other bushes in a variety of localities and is the commonest species of hairstreak in Europe. With wings closed, the butterfly is very effectively camouflaged when resting on green leaves. The upperside is brown but this is practically never displayed.

well as by their present name. One species exists in Europe and a few in Asia but the majority come from South America. These are brightly coloured with brilliant shot blues and reds (*Diorina* and *Ancyluris*), greens (*Lyropteryx*) and other pure reds and blues. An infinite variety of shape and pattern is shown by nemeobiids. It is a most interesting family to study and the life histories of most species are unknown.

Geometridae This is one of the largest and most widespread of the moth families across the world. The average size of a geometer is around 2 centimetres (0·8 inches) or smaller, though a few are larger than this, especially in the tropics. The caterpillars of this family have only two pairs of false legs at the tail end and they walk by arching their backs and drawing the tail end up to the head end. This gives them the name of inchworms in America and, colloquially, loopers. Many of the larvae are stick caterpillars, others live in flowers and take on the colours of the petals on which they live. In Europe there are many species of carpet moths, some of which are found by day. They have a wavy pattern which may be in almost any colour, including some very pretty greens. Carpet moths, as with most geometers, rest with their wings outspread flat either side and those with camouflage patterns press themselves so tightly against their background that they really merge. They may also be seen with their wings held vertically. The pugs are another group with many species. They are quite small, with long, narrow wings that are rounded at the tips. The patterns on some tropical species are very intricate and colourful.

Lasiocampidae The large furry bodied eggar moths comprise this family. They are often coloured in yellow or ochre, tending towards orange-brown. While there are small species, most are medium to large and quite impressive. The moths do not feed and most can be bred with great ease. Lasiocampid larvae feed usually on the leaves of trees or bushes. They are hairy and the hairs on most species are capable of giving a nasty rash if they are handled. The hairs blow in the wind and can cause temporary discomfort to the eyes so care is needed, though they are not in any way dangerous. Some of the tropical Lasiocampidae are coloured in bright greens and oranges. Examples of well-known European species include the Oak Eggar (*Lasiocampa quercus*) which is the largest, the Fox Moth (*Macrothylacia rubi*) whose large furry larvae are found in spring on open moorland, after hibernation, and the Lappet Moth which has probably the best leaf camouflage of any moth – it looks like a bunch of leaves, with a 'nose' that is just like a broken-off stalk!

Saturniidae This family comprises the silkmoths of the world and, although not a saturniid, the silkworm will also be included here. These are the largest and most spectacular of moths and, because they attract so much attention and interest, they will be given greater coverage with some details of breeding the various species in this chapter, and the same will apply in the description of the large hawkmoths (*Sphingidae*) which follows this section.

The Silkworm (*Bombyx mori*) originated in China and the secret of silk and its origin was kept by the Chinese for centuries. The species does not exist in the wild now and the moths have become so domesticated that they have lost the power of flight. They are kept in open trays and there they remain. Similarly, the caterpillars may be kept in the open and not in a cage like other species.

Silkworm eggs should be kept refrigerated for the winter; otherwise they hatch before their foodplant is out in the spring. Mulberry is the only satisfactory foodplant and, although it is often difficult to find a tree nearby, it can be grown quite quickly, especially in a greenhouse. White Mulberry grows quicker than Black Mulberry and it responds well to hard pruning by putting out vigorous shoots. Newly hatched Silkworms are transferred, on the tip of a fine paint brush, to freshly picked leaves in a plastic box. While they are small the closed box is better as it keeps the leaves fresh. Leaves are picked individually, without stems, and laid flat in the box. Larger larvae are kept in open trays and the leaves are put on top of them several times a day. They climb up on top of the leaves and eventually a layer of leaves builds up under the larvae. This is not harmful as long as it

Below
Silkworms are the larvae of the Silkmoth (*Bombyx mori*) which no longer exists in the wild state but originated in China. The silk that they spin as a cocoon is wound off and spun into thread for weaving. Although modern man-made fibres have reduced silk production it is still carried out, especially in the Far East.

Right
The Silkmoths (*Bombyx mori*) have lost the ability to fly and are kept for breeding in open trays which they seldom leave. The pair in the centre are mating. A wild species, *Bombyx mandarina*, exists. It greatly resembles the Silkworm moth but is smoky grey-brown in colour.

is not so moist that mould forms. When Silkworms are about to spin they have a rather translucent appearance. At this stage they need shredded wood fibres, straw or something similar which is placed around the perimeter of the tray and it is here that they spin their cocoon. If it is intended to breed from the cocoons, they are left in place and the moths emerge in a little over a month. They need no special conditions to breed, but are best kept in a paper-lined tray. They lay eggs on the paper which can later be cut up and stored in a plastic box for the winter. If you wish to reel the silk you can do so without boiling the cocoon and killing the pupa inside (which is the normal method). The hot water loosens the silk but, in practice, I have found that the thread comes off quite well, once all the loose silk fluff has been pulled off the outside. Eventually, just one thread keeps coming and this can be wound on to a cotton reel. The more ambitious can spin the silk into thread and weave it and this makes a very interesting school project.

Most of the Saturniidae spin silk, although it is coarser than that of the silkworm. A number of species have been used commercially and research is being carried out on ways of improving the silk and the methods of farming the silkmoths.

Unlike butterflies, which need growing foodplant for egg-laying, flowers for food and plenty of sunshine, the Saturniidae will pair and lay in open netting cages with relatively little pampering. Their mouthparts are either rudimentary or absent so they do not feed. They live for about two weeks, just long enough to reproduce. The females attract males by scent and, before pairing, they hang with the tip of the abdomen projecting to give off the scent. This is known as 'calling' and, if a female is tethered or caged outside, she will attract wild males (assuming that she is in the right locality).

Eye spots or transparent windows feature on practically all saturniids. Most are easily sexed by reference to the antennae which are more pectinate (comb-like) in the male. The female usually has a noticeably fatter body and in many species the male has more hooked (falcate) wings.

Antheraea species display most of the typical characteristics of this family and *A. pernyi* from India and the Far East is a most satisfactory species for the beginner to try to breed. The caterpillars grow to an impressive size and they feed on convenient foods such as willow, oak, apple and hawthorn. Related species include the Tussore Silkmoth (*A. mylitta*) which has a curious egg-like cocoon, *A. roylei* which has much in common with *A. pernyi*, and *A. polyphemus* from America. This last species has silver spangles, like liquid mercury, along the sides of the caterpillar, and the moth has four prominent eye spots and is very variable in its shade of colouring. *Antheraea yamamai* in Japan is one of the few species that spends the winter in the egg stage. The moth is sometimes a very bright yellow though the average form is a little like *A. pernyi*. Green silk is spun by the caterpillar. The Squeaking Silkmoth, also in Japan, is another that has a winter egg. Its fascinating caterpillar (referred to earlier for its counter-shaded colouring) is able to squeak quite loudly if it is disturbed! This species is a must for the breeder as it has a very unusual cocoon (also described earlier) and the moths are sexually dimorphic.

Below
The Spanish Moon Moth (*Graëllsia isabellae*) is atypically exotic for a European species. Sadly its range is very restricted and it is becoming scarcer. It is now protected by law in some countries and may, in the future, survive only if strengthened by stock reproduced in captivity.

Among the most spectacular species are the moon moths, of which *Actias selene* from India is a great favourite. The male is beautifully kite-shaped and coloured in the softest shades of green, yellow and maroon. The female has lovely furled tails. The moths breed easily in captivity. Their larvae are coloured with patches of red and black at first, then they become red all over with black spots and at the third skin change they are green, covered with orange tubercles. The American Moon Moth (*A. luna*) is smaller but very elegantly shaped. Its larvae feed on walnuts. There are also moon moths in the Far East. Sumatra has the rather scarce and exceptionally long-tailed *A. maenas* and in Africa there is the lovely *Argema mimosae* which is a lovely clear lime green colour. The most magnificent of all, *Argema mittrei*, comes from Madagascar. It is a real giant, males having tails approaching 20 centimetres (7·8 inches) long and both sexes are beautifully coloured in canary yellow and orange. It has seldom been bred outside its native country but in America attempts have been reasonably successful using Poison Ivy and other poisonous plants for the larvae.

Europe has rather few Saturniidae. The exotic Spanish

Below
The caterpillars of the Spanish Moon Moth have markings that are a very good imitation of the parts of the pine around the base of the needles where they live. The cocoon is spun among moss on the surface of the ground.

Above
Silk is produced commercially from some of the Saturniidae as well as from the Silkmoth. This Tree of Heaven Silkmoth (*Philosamia cynthia*) is a form found in India. Other forms exist throughout Asia, America and Europe.

Left
Caterpillars of the Saturniidae or giant silkmoths are often very spectacular. The shape of this *Saturnia pyri* is typical of a number of species and there is an infinite variety of decorations and colourful knobs to be found on such larvae. This is Europe's largest species.

Above
Europe has only a limited number of Saturniidae and this Emperor Moth (*Saturnia pavonia*) is the only one found in Britain. Its caterpillars show the relationship to their exotic cousins with colouring that is at least as striking. The cocoon has an open end with bristles pointing conically outwards so that the moth can emerge but nothing can enter while it is resting in hibernation.

Moon Moth (*Graëllsia isabellae*) comes from restricted areas in Spain and the Jura Mountains in France. The largest is *Saturnia pyri* which is like a very large, black and grey Emperor Moth. Larvae of *S. pyri* are very attractive and not unlike species from hotter parts. Unfortunately they are not always easy to rear outside their own habitat. The smaller Emperor Moth is common throughout Europe, especially on heaths where it feeds on heathers. The larvae are black at first and they cluster, only splitting up when quite large. At this stage they are green, ringed with black and decorated with prominent tubercles of pink, mauve or yellow. This is a day-flying species. The very active male is basically orange. The female has the same pattern but in shades of pinkish grey. The Tau Emperor (*Aglia tau*) is not found in Britain but it is widespread in the rest of Europe. The moth is richly coloured in orange, with four prominent blue and black eye spots. The larvae are attractively spined and are a pale lime green colour with markings in maroon. The spines are shed at the last skin change.

Important among the Saturniidae are the atlas and cynthia moths (*Attacus* and *Philosamia*). *Attacus atlas* occurs from India, through the Malay Peninsula to southern China. It is reputed to be the largest moth in the world, though females of the New Guinea *Coscinocera hercules* are as large as *Attacus atlas*. In northern India there is the beautiful, chocolate-coloured

Above
This giant silkmoth, *Attacus atlas*, is found from India eastwards through China to Malaysia. The markings on the wingtips are reminiscent of snakes' heads; these and its sheer size (25–30 centimetres, 10–12 inches across) must give it some protection. The larvae feed on many trees and especially like Tree of Heaven.

Right
Caterpillars of *Automeris io*, the American Bull's Eye Moth, form a colourful cluster of stinging spines that no sensible bird will try and tackle. It is a relative of this which causes a fatal human disease in Venezuela, although *Automeris io* only stings like a nettle.

Attacus edwardsii. Both species can be reared in captivity, feeding on privet, willow, rhododendron and other trees, but they thrive especially well on the Tree of Heaven. The larvae appreciate a warm, humid atmosphere and need ideal, clean conditions. They are white, a colouring that is given by a deposit of white wax all over the body, which is a very pale, translucent green. There are numerous, long, fleshy tubercles, especially on *A. edwardsii* larvae.

Philosamia moths bear the same patterning as *Attacus*, though they are about half the size. Their larvae are also white and similar in general appearance and they have many foodplants in common, especially the Tree of Heaven, which is their principal foodplant and because of which they are often referred to as the Tree of Heaven silkmoths. India has the most beautiful form, *P. cynthia canningii*, which is ginger with lilac bands down the wings. A similar, though less bright, form, *P. cynthia ricini*, also comes from India and this produces eri silk which is used for manufacturing. It is continuously brooded, while the other forms have a period of diapause. *P. cynthia* has forms that have established themselves in Italy and France. These are greenish or khaki and rather similar to the American *P. cynthia advena*.

South America has the genus *Rothschildia* which bears a strong resemblance to *Attacus*. It includes *R. orizaba*, the largest, *R. jorulla* and *R. forbesi*. Their larvae feed on privet and are colourful in shades of apple green with markings of purple or orange. *R. jacobaeae* is a lovely maroon-coloured species and there are others with this colouring.

North America has the robin moths (*Hyalophora*). The most colourful, in shades of charcoal, red and white, is *H. cecropia*. This is a common and widespread species, ranging north well into Canada. The larvae are a beautiful bluish green, decorated with knobs of yellow, red, orange or blue, often with a combination of all these colours. They are easy to rear while young but after the second skin change they die unless they can be given plenty of space and ideal conditions, preferably with growing food. This applies to the other species in this genus. *H. euryalis* (*rubra*), *H. gloveri* and the rare *H. columbia* have a rather similar pattern of wine red and grey. They are variable from region to region and sometimes, therefore, not easy to distinguish without detailed examination.

The female Cherry Moth (*Callosamia promethea*) bears some resemblance to the last genus, in pattern and colour, which is maroon. The male is jet black with greenish ochre borders. The larvae are rather unusual, being pure white with colourful rounded knobs on the thoracic segments. Also in North America, there is a notable *Automeris* species, the Bull's Eye Moth (*A. io*). The male has bright yellow forewings and the female's are orange-brown. On the hindwings both have prominent target-like eye spots.

In fact, *Automeris* should really be considered as a South American genus as there are many dozen species in that region. The general pattern of wing shape and hindwing eye spots is consistent throughout the genus and many can only be identified by a specialist because of their similarity. The larvae are covered with branching spines and are capable of giving a sting like a nettle.

Citheronia and *Eacles* are two related genera which occur mainly in South and Central America but range in to the north as well. They pupate underground and do not spin a silk cocoon. Their wing shape tends to be narrow and elongated and it is probable that in the evolutionary time scale they are between the more normal saturniids and the hawks or sphingids. Their most striking feature is their very exotic horned larvae which are variable in colour and often exceedingly bright. These are known as the 'hickory horned devils'. There are many species in the south but they are best known from the northern *Citheronia regalis* and *Eacles imperialis*.

In Australia and the tropical islands to the north, there are *Antheraea* species which often feed on eucalyptus. Saturniids are not so well represented in this area, nor are they so well known, although new species may well be found in the less accessible parts.

Africa, on the other hand, has an enormous range of Saturniidae. Many do not spin a cocoon but pupate underground. A common species is *Bunaea alcinoë*, large and coloured in chocolate and orange. It can be reared on privet. Unfortunately, the majority of African species do not accept substitute food-plants and little rearing has therefore been done. A light trap set up in the right areas will bring a very varied harvest of *Nudaurelia*, *Pseudobunaea*, *Imbrasia*, *Lobobunaea* (these are large species, often with bright colouring and typical eye spots); species with a similar patterning to the Emperor Moth such as *Heniocha* and *Usta*; species with leaf pattern including *Micragone*, *Goodia* and *Holocerina*; and *Epiphora* which is similar to *Rothschildia* and *Hyalophora* from South and North America respectively.

Below
Argema mittrei, a Madagascan saturniid, is one of the most splendid moths in the world. It quite dwarfs the large *Hibiscus* flower that it is resting on. The cocoon is possibly the largest of any species and is made of a silk that is shining silver in colour.

Sphingidae There is a fascination about the hawkmoths which makes them one of the best known of all groups, though in some countries they are so numerous that, scientifically, we still have a lot to learn about them. The attraction stems partly from the robust, streamlined appearance and partly from the large, boldly patterned caterpillars. The forewing is usually several times the size of the hindwing. At rest, the wings are held either folded back tightly arched over the body, or swept back in a V-shape like a modern aircraft. Some of the species hover and it is from this that the name hawkmoth comes. Most have a very long proboscis or tongue which can probe deeply into the nectaries of flowers with a long corolla. They rely on having plenty of food as they use up much energy in flight and it is important to remember this when breeding them. While some will lay eggs on the cage or other container, many need to lay on growing foodplant like butterflies.

Right
The Death's Head Hawk (*Acherontia atropos*) is famous for the skull marking on the thorax and the fact that the moth can make a loud squeaking noise. It occurs mainly in Africa but it migrates throughout Europe and is one of the most prized species of all.

Below
The hawkmoths or Sphingidae are known for their larvae which generally have a horn at the tail, and the oblique stripes along the sides of this *Manduca sexta* from America are another typical feature. This is one of the most handsome of hawkmoth larvae and it has several different colours in its earlier instars.

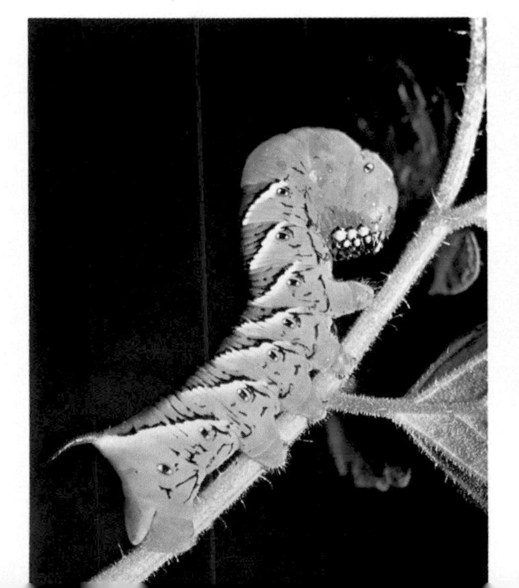

North and South America are very rich in hawks. There are parts of the southern United States and Central America where they are so numerous that in the mornings many hundreds can be seen clustering on walls and lamp-posts, or wherever a light has been shining. It is possible to pick them off simply into a basket, and the moths make no attempt to fly away. It is necessary, however, to get up early as the birds learned this trick long before we did and they get a very substantial breakfast from these moths, some of which are very large indeed. *Pseudosphinx tetrio*, a peppered grey species, has a wingspan of up to 18 centimetres (7 inches) and a body in proportion. Another giant is *Cocytius*, of which there are several species, with dark moss green forewings and with hindwings that are semi-transparent, edged in yellow. *Erynnis* is one of the largest genera here; the species range from quite small ones to others which are three or four times the size. They all have a pattern of orange or yellow transverse stripes, with bands of black, on the body, hindwings matching the stripes and forewings with a camouflage design in grey or brown.

North American hawks include many from further south including *Erynnis*. *Sphinx* is well represented by over a score of grey, rather distinguished looking species related to the European *S. ligustri*. Their larvae, which sit up in a sphinx-like fashion, give the genus its name and in America the name sphinx moth is more usual than hawkmoths. *Manduca*, the tomato and tobacco hornworms, has characters of both the last genera and there are many other groups, some of which are endemic and others, such as the eyed hawks, which are also represented across the world.

Above
One of the most colourful of all sphingid larvae is this Spurge Hawk (*Hyles euphorbiae*) from Europe. It migrates to Britain where it is a prized rarity, though in other regions it is quite common.

Left
Hyles euphorbiae is one of a group of hawks which have a basic shape and patterning that is similar. Their caterpillars are all spectacular and some of the *Hyles* (once known as *Celerio*) will hybridize.

Right
The pink of this Small Elephant Hawk (*Dielephila porcellus*) makes it appear to be conspicuous, but when at rest on suitably coloured flowers it is entirely camouflaged, like the related Large Elephant Hawk (*D. elpenor*).

Africa, too, is rich in hawks. A mercury vapour lamp set up in a good wild area might attract a hundred different species of sphingids at any one season. The range and variety of pattern and colour are tremendous. We can only mention a few without going too deeply in to the subject, as they can become an absorbing study. *Temnora* has many species which are mostly angular in shape and have a camouflage pattern. *Nephele* species are widespread. They have fat, rounded bodies, often coloured in khaki greens or greys, with streamlined swept-back wings. A particular beauty is the bright green *Euchloron megaera*, which reveals wings of bright orange as soon as it takes flight. There are several *Hippotion* species including the well-known Silver-striped Hawk, which is found as a rare migrant in Europe. Another famous rarity in Europe that comes from Africa is the Death's Head Hawk (*Acherontia atropos*). Finding a Death's Head has always been the keen ambition of many collectors. They are usually found as larvae or pupae in potato fields.

In Asia sphingids are also well represented and one of the largest genera that is found only in this region is *Theretra*. These are particularly streamlined in shape and have a very pointed tip to the abdomen, with all their lines and patterning accentuating this. Many are coloured ochre and olive green and some have pink like *Hippotion* which is also represented in this region. *Oxyambulyx* from Malaysia is heavy, yet streamlined–reminding one rather of a heavy V-bomber or troop transport plane!

Various genera of bee hawks are found throughout the world. These have been mentioned in earlier chapters in relation to mimicry, for they are incredibly bee-like and must be left alone by any predator that is not prepared for a risky fight. *Hemaris* is the genus of bee hawks in Europe and there are many more in North America, together with the related *Macroglossa* or hummingbird hawks. In the Far East and in Australia, there is another genus of bee hawk, *Cephonodes*. This has a large green abdomen, banded with maroon, and wings that are transparent between the fine black veins. The caterpillars feed on Gardenia and one of my most memorable finds was when I came across hoards of these all over the bushes in Darwin, north Australia. One of the recurring dreams of the keen 'bug man' is of coming across a super abundance of some of his favourite species, and this was just like that–only real!

A few of the hawkmoths do not have a proboscis and therefore do not feed. European examples include the Poplar Hawk (*Laothoë populi*), the Lime Hawk (*Mimas tiliae*) and the Eyed Hawk (*Smerinthus ocellata*). When breeding these the conditions are therefore greatly simplified and the moths can be left in a plain cage to pair and lay. On hatching, the caterpillars are transferred to rearing boxes or to sleeves outside and they are reared with great ease. The North American *Pachysphinx* can be treated likewise. These moths rather resemble the Poplar Hawk but they are giants, some three or four times the size. All the other eyed hawk genera (*Callasymbolus*, *Paonias* etc.) also have no proboscis.

More care has to be taken to breed the other hawks that do feed. Potted flowers are ideal and if cut flowers are provided they need changing daily. Most will require potted foodplant and respond best to being placed out of doors. The elephant hawks (*Dielephila*), respond well to these conditions and lay their tiny, green, pearl-like eggs on the leaves of willow-herb. The *Hyles* (once known as *Celerio*), which are closely related, should be treated likewise. In Europe these include the Bedstraw Hawk (*Hyles galii*), the Spurge Hawk (*H. euphorbiae*) and the Striped Hawk (*H. livornica*). They all have a great family resemblance and will sometimes hybridize.

Arctiidae This family is best known for the tiger moths which are showy species especially numerous in Europe. The caterpillars of tiger moths, known as woolly bears, are dense hairy larvae, usually not patterned but an overall colour of dark brown.

Right
Arctia caja is a well-known example of the tiger moth family (Arctiidae) which includes some of the most colourful moths known. The caterpillar is the famous 'woolly bear' found throughout Europe, with allied species in all other continents.

The best known is that of the Garden Tiger (*Arctia caja*), which is very handsome both as moth and caterpillar. The Cream-spot Tiger (*A. villica*), which has bright yellow hind-wings, beautifully tinged with pink, is rarer than *A. caja*. The larvae have shorter hairs and are blacker. A day-flying species of great beauty is the Scarlet Tiger (*Callimorpha dominula*) and its larvae are patterned in black and yellow with relatively sparse hairs. The related Jersey Tiger (*Euplagia quadripunctaria*) is another colourful day-flier. This species congregates in huge numbers on the Greek island of Rhodes in the 'Valley of Butterflies'. If someone claps their hands the moths take off, creating a most impressive show of black and orange filling the air. The largest European tiger, *Pericallia matronula*, is not a common species and is highly prized by enthusiasts. It is found only in restricted areas, with a preference for mountains. One of the showiest of all is *Ammobiota festiva*, whose intricate patterning and bright red colouring is especially attractive.

The wings of ermine moths have a basic colouring of white or buff, covered randomly with specks of black. The body is usually coloured yellow. This pattern is adopted by many species across the world but the ermines also cover species with a similar build and quite different colouring. The caterpillars are like woolly bears and feed on many low-growing weeds, seldom with any specific need for a particular plant.

Footman moths are also Arctiidae. They are small, with rounded, narrow wings. They are seldom colourful but the Crimson-speckled Footman (*Utethsia pulchella*), found all over Europe and Asia, is very bright. The larvae are short-haired, and some that feed on lichens are camouflaged to match them exactly.

Noctuidae This is one of the largest families of moths. To the layman they are often very much alike in appearance and, perhaps, rather dull. The differences become more apparent as you study them, and the relationships of the different groups across the world can be very interesting. In America, the larvae are known as cut-worms and there are species that are serious agricultural pests, since when feeding they tend to cut through stems rather than leaves alone, so the whole plant is affected. Several species are used in entomological research because they breed profusely and are ideally suited to laboratory conditions.

There are some very colourful Noctuidae, particularly in the tropics. The outstanding example is *Baorisa hieroglyphica* from Asia, which is white with patches of yellow and red and a very curious streaking in black and blue.

The noctuids so far referred to are mostly small to medium in size and what one might consider as being rather average moths in shape and colour, and are typical of the majority of this family. There are a few notable exceptions which are still classified in this family and two, which are in Asia, display rather effective leaf camouflage. *Othreis* has colourful hindwings which are concealed while at rest, and the much more impressive genus *Phyllodes* has very unusually shaped wings and the different species show a range of bright colours. In America there are noctuids which have a similar wing shape to the saturniids, such as *Ascalapha odora* which is charcoal-coloured and rather beautifully marked with intricate wavy lines. *Thysania zenobia* is a pale grey version of this and *T. agrippina*, which can have a wingspan as great as that of *Attacus atlas*, is a scaled-up version of *T. zenobia*.

One further group of noctuids which must be mentioned is the underwings (*Catocala*). There are several in Europe. They are triangular in shape at rest, medium-sized and patterned in grey with wavy lines which camouflage them well. With wings open they display vivid hindwings, blue with black bars in the case of *C. fraxini* and scarlet or yellow in the other species. North America has many species of *Catocala*. Their larvae are rather twig-like and press themselves hard against the branches by day, so that they are practically invisible.

Other moths Without devoting a special section to each of the other prominent groups of moths there are some which deserve mention, though the number of families throughout the world is so great that we must be selective.

The Lymantriidae are noted for their colourful, hairy larvae. The Vapourer (*Orgyia antiqua*), which has equivalents in America, has been mentioned in an earlier chapter for its attractive caterpillar and the wingless female which never moves from the cocoon, even laying her eggs on it. The pale and dark tussocks (*Dasychira*) likewise have highly decorated hairy larvae. The larvae of the gold and brown tails are striking and the female moths use the coloured tuft at the tail as a protection for the

Above left
Caterpillars of the tussock family, Lymantriidae, are some of the most colourful of all. They have beautiful tufts and brushes of hair. This *Dasychira abietis* is an uncommon species from Bavaria.

Above
The Hepialidae, ghost and swift moths, is a family of moths which evolved comparatively early. They have a characteristic wing shape and giant versions come from the tropics. The larvae of this female *Hepialus humuli* live on roots underground; the male is the pure white Ghost Moth.

eggs which they cover with adhesive and then dust with the tail! As previously mentioned, the hairs of the caterpillars in this group can cause a skin rash.

The Notodontidae or prominents are found in all geographical regions. They include the puss moths (*Cerura*), the kitten moths and the Lobster Moth. The moths are small to medium in size and some of the exotic species are brightly coloured. The caterpillars display unusual features and many have shapes and markings, which make them hardly recognizable as caterpillars. They are not camouflaged, but have a conspicuous disruptive coloration.

The family Ctenuchidae (formerly Syntomidae) has very wasp-like moths with narrow wings and slender bodies. Some are such good mimics that they deceive even an entomologist. Others are rather similar to the burnet moths (Zygaenidae) and a common European example is *Syntomis phegea*, the black and white Polka Dot Moth. The Zygaenidae comprises mainly the burnet moths of the downs and grasslands; foresters, which are shaped like burnets but coloured metallic green all over; and the erasmias of tropical Asia. *Erasmia pulchella* is many times larger than the other zygaenids with a unique wing shape. It is white, coloured with iridescent green, blue and yellow. *E. sanguiflua* is bigger still and coloured deep blue and purple with heavy veining in black. Many other species also exist in Sri Lanka, India and Malaysia.

Above right
The Zygaenidae are colourful day-flying moths found mainly in Europe though they are also represented in Africa and other tropical parts. In Europe they are often a feature of grassy hillsides and mountains where they cluster on flowerheads so densely that the flower is invisible!

Right
The Puss Moth (*Cerura vinula*) is particularly well known for its spectacular caterpillars. The moth is rather delicately marked and has a lovely fluffy texture which undoubtedly accounts for its vernacular name.

Perhaps the most colourful moth in the world is *Urania ripheus* from Madagascar. It is more like a swallowtail butterfly than a moth, with white-fringed multitails to the hindwings and magnificently patterned in metallic greens, black and fiery orange. Sometimes known as the Sunset Moth, this species has one close relative on the African mainland which is rather scarce, *Urania croesus*. In South America *U. leilus* and *U. fulgens* are found. They are obviously similar but lack the fiery oranges and reds.

Australia has some highly colourful moths in the family Agaristidae. *Agarista agricola* is not uncommon and is marked with patches of turquoise, red and yellow, with black and green in a most delightful combination. Further north, in New Guinea, there is a family of moths that are exceptionally brightly coloured, which might possibly be taken for Agaristidae. These are *Milionia* which in fact are classified as Geometridae although their shape and build would not suggest this at all. Although they attract much attention because of their exceptional brightness, very little is known of their life history.

Below
One of Europe's most familiar sphingids is the Poplar Hawk (*Laothoë populi*) which has a range right across to Japan and counterparts in America. It breeds easily, feeding on poplars and willows and there are several colour forms of the larvae.

Breeding butterflies and moths

Above
This large swallowtail, *Papilio aegeus*, is one of the species that can be bred in captivity relatively easily. Their foodplant *Citrus* can be substituted with the hardy shrub *Choisya ternata*. Plenty of flowers are needed and a warm, light place such as a greenhouse.

Right
One of the best swallowtails to breed is *Papilio machaon* which is found in many parts of the world. In England a subspecies, *P. m. britannicus*, has evolved in isolated parts of Norfolk which only likes fen country, whereas in other countries it tends to inhabit dry and mountainous country.

One of the best ways of getting to know Lepidoptera and their habits is to keep and watch them at home, preferably in as natural surroundings as possible. There is also great satisfaction in watching a butterfly or moth emerge that you have nurtured from the egg, through the larval stage to the chrysalis. Lepidoptera multiply in captivity more than they possibly can in the wild, so breeding butterflies for release can be very worthwhile in aiding conservation.

This chapter is designed to help would-be breeders by outlining methods used at the present time, but undoubtedly new methods will continue to be found. There are no hard and fast rules, and a good deal relies on trial and error mixed with common sense. It will be apparent from earlier chapters that many moths are easier to breed from than butterflies because they do not need to be set up with foodplant and flowers in sunshine, but rearing from eggs or larvae of moths or butterflies is much the same.

It is generally agreed that it is more satisfactory to rear larvae on growing plants rather than in breeding boxes but as the right plant may not always be prepared in advance or other circumstances may dictate the use of rearing boxes, it is a good thing to know about this method.

You may have been fortunate enough to find some eggs or they may have been obtained from a breeder. They are likely to be loose or on small fragments of withered foodplant, so they are best kept in a small, transparent, plastic box. Keep them out of direct sunlight as this is harmful. The eggs may well darken just before hatching. Let the newly hatched larvae feed on their egg-shells for a while and then transfer them to fresh food. This can either be done by picking them up gently on the tip of a soft paint brush or by putting in a small amount of their foodplant (for about an hour) and transferring this when they have crawled on to it. If you have a pot of the correct growing foodplant available it would be as well to transfer the young larvae straight on to this. If not, then use a larger plastic box (the size of a sandwich box), line it with paper and put freshly cut food on top of the paper lining. The box should be of clear plastic so that you can see everything inside.

The paint brush method of transferring larvae is gentle, but avoid the temptation to sweep the larvae in quantity as this may harm them unless they are one of the kinds that curl up into a ball. Select foodplant that is dry, fresh and clean and see that it is free from aphids' honeydew. Take a reasonable length and arrange it on top of the paper lining in the box, in a bowed fashion, so that it is not lying on the bottom but is accessible to larvae even if they wander all round the box. Similarly, do not put an odd leaf or two on the bottom and hope the larvae will

Left
The Scarlet Tiger Moth (*Callimorpha dominula*) is one of the most beautiful European moths. It flies by day in June. The caterpillars can be found often along river banks, usually in colonies but they do not live in clusters. The best time to look is in spring.

Below
From North America comes *Papilio polyxenes asterias* which is related to *P. machaon* and will breed with it. The larvae also feed on fennel and other umbellifers. The eggs are distinctly smaller than those of *P. machaon* but the larvae are not easily distinguished. In the adult there is considerable colour variation, some being so dark that they are practically all black, while the lightest of males could almost be taken for *P. machaon*.

not stray from it. It is possible to cram too much food into a plastic box so be moderate.

Cleaning out must be done daily, renewing the food at the same time, and the following procedure should always be used. Put a new liner in the box; the larvae go in next but do not remove them from the leaves and twigs on which they are resting; cut round them, removing as much old food as possible; and finally put the fresh food in *on top* of the larvae which will crawl up to it (they will not always make their way down to new food if you put them in on top of it). Make sure the boxes are not kept in the sun at all.

Plastic boxes enable you to control the larvae; they cannot wander far from the food and starve and you can see what is happening all the time. Air holes are not necessary as more air is trapped inside than the larvae will need and the closed container keeps the foodplant absolutely fresh.

Eggs that have been laid on growing potted foodplant can be kept in sunshine if necessary and it is all right to use a greenhouse as long as it is well ventilated and the temperature is not allowed to rise much above 35°C. The larvae are left to feed naturally and will need little attention but it is advisable to cover the plant with a bag of fine netting to keep out other predatory insects, woodlice and so on. Most butterfly eggs hatch within a few days or up to three weeks of being laid. Some moth species take a little longer. The eggs of overwintering species would normally be laid in the late summer. In the wild state they would remain in position on the twig, bark or wherever they were laid, right through the winter, exposed to all weathers, until they hatched in the spring. If you have some of these they keep quite well in a plastic box in a very cool place, even in a refrigerator. Bring out the eggs in the spring and watch them carefully, every day, for hatching. Some newly hatched larvae, especially those of *Thecla* and *Lycaena* are minute and might easily be overlooked. If, by misfortune, your eggs hatch before the buds of their foodplant are open, as a temporary measure you can dissect some buds and feed the tiny pieces of leaf to the larvae. The larvae will sometimes even burrow into the bud.

Above
The American Gulf Fritillary (*Dione vanillae*) can be bred in a greenhouse with passionflowers. They need flowers and do well in drier conditions than most butterflies.

Larvae that have been reared in a plastic box and have grown well can be transferred to a breeding cage. A cylindrical cage is very successful and the larvae inside are easily observed. Ventilation is through the lid. As with plastic boxes this type of cage is not suitable for a sunny position because condensation builds up on the sides and the cage becomes too wet. Although more expensive, the best type of cage is made with a wooden frame covered with fine netting. This is not too difficult to construct for yourself if necessary. Not only do larvae do well in this type of netting cage but it is also ideal for pupae and adults. Emerging adults can get a good foothold and climb the netting to dry their wings. The cage can be placed near a window without fear of the sun causing condensation. In the cage, potted food is more satisfactory than cut food. Not only is it better for the larvae but it is also less troublesome because the food requires changing far less frequently. If cut food is used, however, stand it in a jar of water, against one side, so that larvae can crawl up again if they fall. The foodplant ought to be changed once every three or four days even if it still looks fresh and is not finished. If you stand a pot of fresh food beside the old, most of the larvae will transfer themselves. The The next day the old food can be removed and the few remaining larvae clipped off by hand.

The cage should be cleaned out daily if possible, removing any dead or obviously unhealthy larvae, taking care to leave those which are motionless because they are changing their skins. They will be attached to silk pads and must not be pulled off.

Sleeving is a method which can be very successful if intelligently applied. A sleeve is a bag of fine cheesecloth, organdie or mosquito netting, according to the requirement. A sleeve over a potted foodplant makes a very economical cage and it can be used indoors or out. For very small larvae on a young, tender

Above
The Indian Golden Emperor Moth (*Loëpa katinka*) is one of the brightest of all silkmoths. They breed well if left to emerge communally, pair and lay on the sides of the emerging cage. The caterpillars are very showy and feed on grapevines.

Left
The silkmoth that is easiest of all to rear, and much recommended for beginners, is this *Antheraea pernyi* from India and China. The caterpillars are black at first, then change to green and they grow at a tremendous rate, feeding on a number of different trees including oak, willow and beech.

Right
This group of North American Bull's Eye Moths (*Automeris io*) makes a very spectacular show when they open their wings to expose their target-like eye spots. They breed comparatively easily on many different kinds of tree.

plant, the lightweight, soft organdie is best. Indoors, a black open netting sleeve is best, as long as the larvae are not too small, as you can see into it. Outside, a whole tree or branch can be sleeved, with the larvae free inside, living on the fresh growing plant as in the wild state. Black netting sleeves give better air circulation and do not get so damp, but birds can peck through these and a closer-weaved white bag is often safer. Double sleeving can prevent birds from getting in but not always. In a spell of really wet and cold weather, larvae do not do well in sleeves and they are better brought indoors. Never crowd larvae unless they are gregarious, as this often leads to disease. The foodplant must never become stripped of its leaves because if the larvae suffer even partial starvation this weakens them. They will almost certainly die at a later date by catching a disease.

Diseases caused by viruses and bacteria are a problem in rearing Lepidoptera and very little is known about them. The larva hangs limply or lies in a pool of liquid. Sometimes the faeces become very liquid. The causes are generally over-crowding, damp or unclean conditions, wet food, starvation or other factors which weaken the caterpillar. Disease exists even in wild stock, and is sometimes there in latent form ready to manifest itself in adverse conditions. It can seldom be cured and spreads easily. Weak sodium hydroxide solution kills viruses and should be used for washing out cages and boxes either as a precaution or after an outbreak has occurred.

In captivity parasites are seldom a problem to the breeder. Parasites are either wasp-like Hymenoptera or dipterous flies. They lay their eggs either on the skin of a caterpillar or inject eggs into the body with a sharp ovipositor. The eggs develop and produce either a multitude of tiny insects or maybe only one. The result is certain death to the caterpillar. When breeding vanessids and papilionids, freshly formed pupae must not be kept out of doors as they are vulnerable to attack from tiny braconid wasps.

Some preparation for pupating has to be made. Larvae being reared in a cage may pupate hanging from the roof among the foodplant or they may be a species that makes a chrysalis at the bottom of the cage. Moth larvae may need a layer of peat or soil to pupate in at the bottom. Freshly formed pupae should not be disturbed as they are very soft. It is not a good thing to remove even hardened pupae that are hanging from silk pads unless you are experienced. They easily break, leaving the tail segments behind and, in any case, the pads give the necessary anchorage when

Left
The colouring of this Indian Silkmoth (*Cricula andrei*) is very variable but always in lovely shades of leaf-like autumn colour. *C. andrei* can be bred in captivity on plum and related trees.

Right
Spectacular among all moths is the Indian Moon Moth (*Actias selene*). It is a great favourite with breeders as the caterpillars, too, are exceptionally beautiful and it is easy to breed, feeding on rhododendron, willow, apple and other common trees.

Below
Brahmaea japonica is another favourite with breeders. This Japanese moth emerges in winter, producing the first larvae of spring. Although not colourful the pattern and subtle shading of the moth are extremely attractive.

Below right
Brahmaea japonica caterpillars have long black horns both behind the head and at the tail. These are shed at the last skin change and the caterpillar becomes very large. It pupates underground. The larvae feed on privet and are very easy to rear.

the adult emerges. Some sleeved larvae can be left to pupate in the sleeve but generally it is better to remove them as soon as they show signs of wanting to pupate. For underground pupae, make up a tray of peat or soil which can be covered so they do not escape. Butterfly pupae that are suspended or attached by a girdle can be put into shoe boxes where they are safe from parasites and cannot escape. Larvae that are ready to pupate have a very strong instinct to wander and find distant hidden places. Do not be too surprised if they find their way out of your containers; they have to be very secure indeed!

Storage of pupae during the winter must ensure that moisture is retained and that the pupae are cool. This is easily achieved by keeping them in plastic boxes which prevent evaporation (no soft padding is necessary, just the pupae loose in the box) in a very cool outhouse or a refrigerator. There is no fear of them becoming too cold. Although it is better not to remove pupae from their cages normally, for winter storage this is necessary. Make sure you know the approximate time the adults normally emerge and lay the pupae out about a month before this.

Some of the easiest butterflies to rear include vanessids such as the Peacock, *Precis*, satyrids (after hibernation), certain papilios such as *Papilio machaon* or *P. demoleus* and *Pieris* species. Among the moths, the outstanding species are the saturniid, *Antheraea pernyi*, and others mentioned in the section on this family, hawks of the genera *Laothoë*, *Pachysphinx*, *Smerinthus* and *Sphinx*. The Lymantriidae are suitable (remembering the care necessary with the hairy larvae) and the Notodontidae.

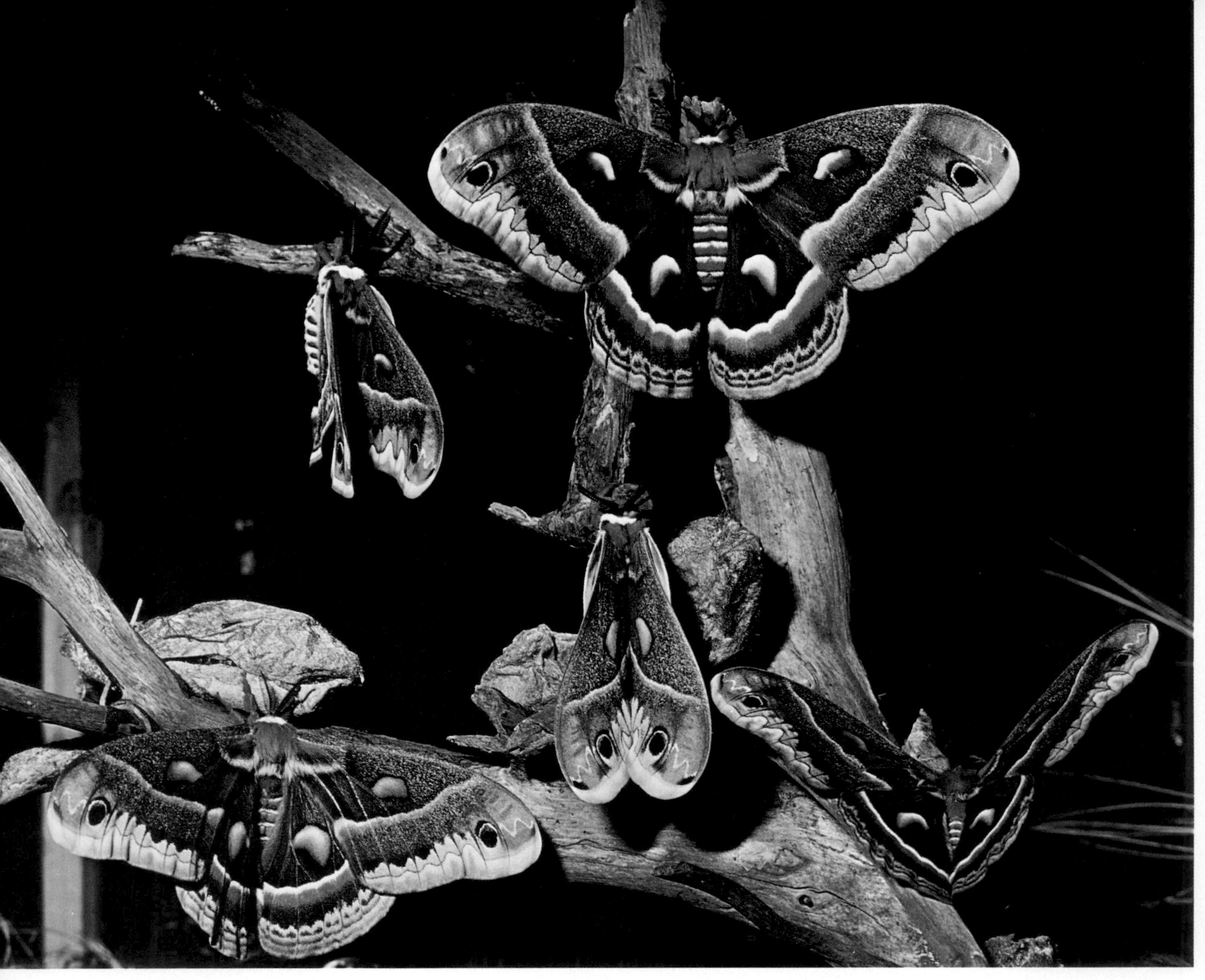

Above
This group of Robin Moths (*Hyalophora cecropia*) from North America illustrates the beauty of the big silkmoths and the fun there is in breeding them. Some skill is needed in raising this species but it does well on growing foodplant out of doors if the weather is fine and not too cold.

After you have reared your Lepidoptera, the next stage, particularly with butterflies, is more difficult. A few methods and tips outlined here may help you and a lot will depend on your own experimenting and ingenuity. In captivity, butterflies need to be tempted into pairing and egg laying by having the larval foodplant and flowers so arranged that both are about 7 centimetres (2·8 inches) from the cage top and spaced so that the butterflies cannot help bumping into both wherever they fly. This is simply a general principle and certain species will prove to be exceptions. Butterflies can be sleeved over a potted foodplant to pair and lay but in an ordinary sleeve they tend to get caught in the folds and trapped until they eventually die. A modified sleeve is necessary, with a flat, circular top. Suspend the cage over the plant and make sure that the leaves are close to the top, together with fresh flowers. Sometimes butterflies will prefer one situation to another with no obvious reason but if you find that one particular spot in the garden or greenhouse suits a species best, take advantage of it and go on using this position.

It seldom helps to provide a large space for egg laying and often the females are induced to lay better in a confined space, where the presence of their foodplant stimulates them to keep laying. A female may lay her batch all at once or a few eggs every so often over a week or two. She may show a preference for a particular shoot or flowerhead and, if this is observed, be careful not to move this piece of foodplant as she may smother it with eggs, but hardly lay elsewhere. Make sure the females are well fed and if you think they are not feeding well enough naturally, place them on the fresh flowers by hand. You will see the proboscis probing the flower if it is feeding. If not, take a long needle and uncurl

the proboscis gently until it starts feeding. You can also feed with honey pads. To make these take a jam jar and pour in honey to a depth of about 0·5 centimetres (0·2 inches). Fill it up with warm water and shake well to mix. Dip in the cotton wool and squeeze out the surplus. Replenish daily or simply moisten the pads, as the sweetness remains in the pad even when the water has evaporated. The pad is placed outside the cage, resting on the netting, and the butterflies (not all species will feed from pads) probe through the netting. To hand feed on honey pads, hold the wings gently but firmly and uncurl the proboscis so that the tip is on the pad. Let the legs rest on the pad as well because they are more at ease like this and they taste through the feet.

By experimenting you will find methods for yourself and with a great many tropical species this is necessary, because little or nothing is known of their life histories. There is nothing to stop anyone with the right plants from rearing exciting, tropical species in temperate countries. It is not easy to obtain livestock of many tropical species but some are available from tropical breeders or perhaps you might have a friend in the tropics. *Passiflora* grows easily and is the food of certain tropical Nymphalidae. Many tropical *Papilio* species will feed on *Citrus* plants, so these too are useful to grow.

Above left
Different kinds of caterpillars will usually live in harmony with each other and these two are being reared in a new experimental tunnel cage placed over coppiced foodplant. The green larva is that of the Poplar Hawk and the white silkmoth caterpillar, feeding on the same willow bush, is *Philosamia cynthia*.

Left
The caterpillar of the Elephant Hawkmoth (*Dielephila elpenor*) gives the moth its name from the way it can extend the front segments into a trunk. The eye-like markings are startling to a predator and if disturbed the caterpillar swells up its front segments, almost like a cobra snake, and writhes violently.

Above
This pairing of the Oleander Hawkmoth (*Daphnis nerii*) is one of only very few ever achieved in captivity. Eggs were obtained from Persia and generation after generation bred on Oleander and periwinkle. This species lives mainly in Africa but migrates rarely northwards to Europe where it is much prized.

Index

Page numbers in italic refer to illustrations